The Evolution of Man, V.1.

Ernst Haeckel

Table of Contents

The Evolution of Man, V.1...1
   Ernst Haeckel..1
   PREFACE...7
HAECKEL'S CLASSIFICATION OF THE ANIMAL WORLD...............................13
THE EVOLUTION OF MAN..17
   CHAPTER 1.1. THE FUNDAMENTAL LAW OF ORGANIC
   EVOLUTION...17
   CHAPTER 1.2. THE OLDER EMBRYOLOGY...30
   CHAPTER 1.3. MODERN EMBRYOLOGY..39
   CHAPTER 1.4. THE OLDER PHYLOGENY...52
   CHAPTER 1.5. THE MODERN SCIENCE OF EVOLUTION........................60
   CHAPTER 1.6. THE OVUM AND THE AMOEBA..73
   CHAPTER 1.7. CONCEPTION...91
   CHAPTER 1.8. THE GASTRAEA THEORY...102
   CHAPTER 1.9. THE GASTRULATION OF THE VERTEBRATE.*.............119
   CHAPTER 1.10. THE COELOM THEORY...141
   CHAPTER 1.11. THE VERTEBRATE CHARACTER OF MAN...................156
   CHAPTER 1.12. EMBRYONIC SHIELD AND GERMINATIVE AREA......174
   CHAPTER 1.13. DORSAL BODY AND VENTRAL BODY..........................186
   CHAPTER 1.14. THE ARTICULATION OF THE BODY.*..........................201
   CHAPTER 1.15. FOETAL MEMBRANES AND CIRCULATION................217

The Evolution of Man, V.1.

Ernst Haeckel

- PREFACE.
- HAECKEL'S CLASSIFICATION OF THE ANIMAL WORLD.
- THE EVOLUTION OF MAN.

    - CHAPTER 1.1. THE FUNDAMENTAL LAW OF ORGANIC EVOLUTION.
    - CHAPTER 1.2. THE OLDER EMBRYOLOGY.
    - CHAPTER 1.3. MODERN EMBRYOLOGY.
    - CHAPTER 1.4. THE OLDER PHYLOGENY.
    - CHAPTER 1.5. THE MODERN SCIENCE OF EVOLUTION.
    - CHAPTER 1.6. THE OVUM AND THE AMOEBA.
    - CHAPTER 1.7. CONCEPTION.
    - CHAPTER 1.8. THE GASTRAEA THEORY.
    - CHAPTER 1.9. THE GASTRULATION OF THE VERTEBRATE.*
    - CHAPTER 1.10. THE COELOM THEORY.
    - CHAPTER 1.11. THE VERTEBRATE CHARACTER OF MAN.
    - CHAPTER 1.12. EMBRYONIC SHIELD AND GERMINATIVE AREA.
    - CHAPTER 1.13. DORSAL BODY AND VENTRAL BODY.
    - CHAPTER 1.14. THE ARTICULATION OF THE BODY.*
    - CHAPTER 1.15. FOETAL MEMBRANES AND CIRCULATION.

The Evolution of Man, V.1.

TRANSLATED FROM THE FIFTH (ENLARGED) EDITION BY JOSEPH MCCABE.

THE EVOLUTION OF MAN

A POPULAR SCIENTIFIC STUDY

BY

ERNST HAECKEL

VOLUME 1.

HUMAN EMBRYOLOGY OR ONTOGENY.

GLOSSARY.

ACRANIA: animals without skull (cranium).

ANTHROPOGENY: the evolution (genesis) of man (anthropos).

ANTHROPOLOGY: the science of man.

ARCHI–: (in compounds) the first or typical—as, archi–cytula, archi–gastrula, etc.

BIOGENY: the science of the genesis of life (bios).

BLAST–: (in compounds) pertaining to the early embryo (blastos = a bud); hence:—
Blastoderm: skin (derma) or enclosing layer of the embryo.
Blastosphere: the embryo in the hollow sphere stage.
Blastula: same as preceding.
Epiblast: the outer layer of the embryo (ectoderm).
Hypoblast: the inner layer of the embryo (endoderm).

BRANCHIAL: pertaining to the gills (branchia).

CARYO–: (in compounds) pertaining to the nucleus (caryon); hence:—
Caryokineses: the movement of the nucleus.
Caryolysis: dissolution of the nucleus.
Caryoplasm: the matter of the nucleus.

CENTROLECITHAL: see under LECITH–.

CHORDARIA and CHORDONIA: animals with a dorsal chord or back–bone.

COELOM or COELOMA: the body–cavity in the embryo; hence:—
Coelenterata: animals without a body–cavity.
Coelomaria: animals with a body–cavity.
Coelomation: formation of the body–cavity.

CYTO–: (in compounds) pertaining to the cell (cytos); hence:—
Cytoblast: the nucleus of the cell.
Cytodes: cell–like bodies, imperfect cells.
Cytoplasm: the matter of the body of the cell.
Cytosoma: the body (soma) of the cell.

CRYPTORCHISM: abnormal retention of the testicles in the body.

DEUTOPLASM: see PLASM.

DUALISM: the belief in the existence of two entirely distinct
principles (such as matter and spirit).

DYSTELEOLOGY: the science of those features in organisms which refute
the "design–argument."

ECTODERM: the outer (ekto) layer of the embryo.

ENTODERM: the inner (ento) layer of the embryo.

EPIDERM: the outer layer of the skin.

EPIGENESIS: the theory of gradual development of organs in the embryo.

EPIPHYSIS: the third or central eye in the early vertebrates.

EPISOMA: see SOMA.

EPITHELIA: tissues covering the surface of parts of the body (such as the mouth, etc.)

GONADS: the sexual glands.

GONOCHORISM: separation of the male and female sexes.

GONOTOMES: sections of the sexual glands.

GYNECOMAST: a male with the breasts (masta) of a woman (gyne).

HEPATIC: pertaining to the liver (hepar).

HOLOBLASTIC: embryos in which the animal and vegetal cells divide equally (holon = whole).

HYPERMASTISM: the possession of more than the normal breasts (masta).

HYPOBRANCHIAL: underneath (hypo) the gills.

HYPOPHYSIS: sensitive–offshoot from the brain in the vertebrate.

HYPOSOMA: see SOMA.

LECITH–: pertaining to the yelk (lecithus); hence:—
Centrolecithal: eggs with the yelk in the centre.
Lecithoma: the yelk–sac.
Telolecithal: eggs with the yelk at one end.

MEROBLASTIC: cleaving in part (meron) only.

META–: (in compounds) the "after" or secondary stage; hence:—
Metagaster: the secondary or permanent gut (gaster).
Metaplasm: secondary or differentiated plasm.
Metastoma: the secondary or permanent mouth (stoma).
Metazoa: the higher or later animals, made up of many cells.
Metovum: the mature or advanced ovum.

METAMERA: the segments into which the embryo breaks up.

METAMERISM: the segmentation of the embryo.

MONERA: the most primitive of the unicellular organisms.

MONISM: belief in the fundamental unity of all things.

MORPHOLOGY: the science of organic forms (generally equivalent to anatomy).

MYOTOMES: segments into which the muscles break up.

NEPHRA: the kidneys; hence:—
Nephridia: the rudimentary kidney–organs.
Nephrotomes: the segments of the developing kidneys.

ONTOGENY: the science of the development of the individual (generally equivalent to embryology).

PERIGENESIS: the genesis of the movements in the vital particles.

PHAGOCYTES: cells that absorb food (phagein = to eat).

PHYLOGENY: the science of the evolution of species (phyla).

PLANOCYTES: cells that move about (planein).

PLASM: the colloid or jelly–like matter of which organisms are composed; hence:—
Caryoplasm: the matter of the nucleus (caryon).
Cytoplasm: the matter of the body of the cell.
Deutoplasm: secondary or differentiated plasm.
Metaplasm: secondary or differentiated plasm.
Protoplasm: primitive or undifferentiated plasm.

PLASSON: the simplest form of plasm.

PLASTIDULES: small particles of plasm.

POLYSPERMISM: the penetration of more than one sperm–cell into the ovum.

PRO–or PROT: (in compounds) the earlier form (opposed to META); hence:—
Prochorion: the first form of the chorion.
Progaster: the first or primitive stomach.
Pronephridia: the earlier form of the kidneys.
Prorenal: the earlier form of the kidneys.
Prostoma: the first or primitive mouth.
Protists: the earliest or unicellular organisms.
Provertebrae: the earliest phase of the vertebrae.
Protophyta: the primitive or unicellular plants.
Protoplasm: undifferentiated plasm.
Protozoa: the primitive or unicellular animals.

RENAL: pertaining to the kidneys (renes).

SCATULATION: packing or boxing–up (scatula = a box).

SCLEROTOMES: segments into which the primitive skeleton falls.

SOMA: the body; hence:—
Cytosoma: the body of the cell (cytos).
Episoma: the upper or back–half of the embryonic body.
Somites: segments of the embryonic body.

Hyposoma: the under or belly–half of the embryonic body.

TELEOLOGY: the belief in design and purpose (telos) in nature.

TELOLECITHAL: see LECITH–.

UMBILICAL: pertaining to the navel (umbilicus).

VITELLINE: pertaining to the yelk (vitellus).

***

PREFACE.

[BY JOSEPH MCCABE.]

The work which we now place within the reach of every reader of the English tongue is one of the finest productions of its distinguished author. The first edition appeared in 1874. At that time the conviction of man's natural evolution was even less advanced in Germany than in England, and the work raised a storm of controversy. Theologians—forgetting the commonest facts of our individual development—spoke with the most profound disdain of the theory that a Luther or a Goethe could be the outcome of development from a tiny speck of protoplasm. The work, one of the most distinguished of them said, was "a fleck of shame on the escutcheon of Germany." To–day its conclusion is accepted by influential clerics, such as the Dean of Westminster, and by almost every biologist and anthropologist of distinction in Europe. Evolution is not a laboriously reached conclusion, but a guiding truth, in biological literature to–day.

There was ample evidence to substantiate the conclusion even in the first edition of the book. But fresh facts have come to light in each decade, always enforcing the general truth of man's evolution, and at times making clearer the line of development. Professor Haeckel embodied these in successive editions of his work. In the fifth edition, of which this is a translation, reference will be found to the very latest facts bearing on the evolution of man, such as the discovery of the remarkable effect of mixing human blood with that of the anthropoid ape. Moreover, the ample series of illustrations has been considerably improved and enlarged; there is no scientific work published, at a price remotely approaching that of the present edition, with so abundant and excellent a supply of illustrations. When it was issued in Germany, a few years ago, a distinguished biologist wrote in the Frankfurter Zeitung that it would secure immortality for its author, the most notable critic of the idea of immortality. And the Daily Telegraph reviewer described the English version as a "handsome edition of Haeckel's monumental work," and "an issue worthy of the subject and the author."

The influence of such a work, one of the most constructive that Haeckel has ever written, should extend to more than the few hundred readers who are able to purchase the expensive volumes of the original issue. Few pages in the story of science are more arresting and generally instructive than this great picture of "mankind in the making." The horizon of the mind is healthily expanded as we follow the search–light of science down the vast avenues of past time, and gaze on the uncouth forms that enter into, or illustrate, the line of our ancestry. And if the imagination recoils from the strange and remote figures that are lit up by our search–light, and hesitates to accept them as ancestral forms, science draws aside another veil and reveals another picture to us. It shows us that each of us passes, in our embryonic development, through a series of forms hardly less uncouth and unfamiliar. Nay, it traces a parallel between the two series of forms. It shows us man beginning his existence, in the ovary of the female infant, as a minute and simple speck of jelly–like plasm. It shows us (from analogy) the fertilised ovum breaking into a cluster of cohering cells, and folding and curving, until the limb–less, head–less, long–tailed foetus looks like a worm–shaped body. It then points out how gill–slits and corresponding blood–vessels appear, as in a lowly fish, and the fin–like extremities bud out and grow into limbs, and so on; until, after a very clear ape–stage, the definite human form emerges from the series of transformations.

It is with this embryological evidence for our evolution that the present volume is concerned. There are illustrations in the work that will make the point clear at a glance. Possibly TOO clear; for the simplicity of the idea and the eagerness to apply it at every point have carried many, who borrow hastily from Haeckel, out of their scientific depth. Haeckel has never shared their errors, nor encouraged their superficiality. He insists from the outset that a complete parallel could not possibly be expected. Embryonic life itself is subject to evolution. Though there is a general and substantial law—as most of our English and American authorities admit—that the embryonic series of forms recalls the ancestral series of forms, the parallel is blurred throughout and often distorted. It is not the obvious resemblance of the embryos of different animals, and their general similarity to our extinct ancestors in this or that organ, on which we must rest our case. A careful study must be made of the various stages through which all embryos pass, and an effort made to prove their real identity and therefore genealogical relation.

This is a task of great subtlety and delicacy. Many scientists have worked at it together with Professor Haeckel—I need only name our own Professor Balfour and Professor Ray Lankester—and the scheme is fairly complete. But the general reader must not expect that even so clear a writer as Haeckel can describe these intricate processes without demanding his very careful attention. Most of the chapters in the present volume (and the second volume will be less difficult) are easily intelligible to all; but there are points at which the line of argument is necessarily subtle and complex. In the hope that most readers will be induced to master even these more difficult chapters, I will give an outline of the characteristic argument of the work. Haeckel's distinctive services in regard to man's evolution have been:

1. The construction of a complete ancestral tree, though, of course, some of the stages in it are purely conjectural, and not final.

2. The tracing of the remarkable reproduction of ancestral forms in the embryonic development of the individual. Naturally, he has not worked alone in either department.

The second volume of this work will embody the first of these two achievements; the present one is mainly concerned with the latter. It will be useful for the reader to have a synopsis of the argument and an explanation of some of the chief terms invented or employed by the author.

The main theme of the work is that, in the course of their embryonic development, all animals, including man, pass roughly and rapidly through a series of forms which represents the succession of their ancestors in the past. After a severe and extensive study of embryonic phenomena, Haeckel has drawn up a "law" (in the ordinary scientific sense) to this effect, and has called it "the biogenetic law," or the chief law relating to the evolution (genesis) of life (bios). This law is widely and increasingly accepted by embryologists and zoologists. It is enough to quote a recent declaration of the great American zoologist, President D. Starr Jordan: "It is, of course, true that the life–history of the individual is an epitome of the life–history of the race"; while a distinguished German zoologist (Sarasin) has described it as being of the same use to the biologist as spectrum analysis is to the astronomer.

But the reproduction of ancestral forms in the course of the embryonic development is by no means always clear, or even always present. Many of the embryonic phases do not recall ancestral stages at all. They may have done so originally, but we must remember that the embryonic life itself has been subject to adaptive changes for millions of years. All this is clearly explained by Professor Haeckel. For the moment, I would impress on the reader the vital importance of fixing the distinction from the start. He must thoroughly familiarise himself with the meaning of five terms.

BIOGENY is the development of life in general (both in the individual and the species), or the sciences describing it.

ONTOGENY is the development (embryonic and post–embryonic) of the individual (on), or the science describing it.

PHYLOGENY is the development of the race or stem (phulon), or the science describing it.

Roughly, ontogeny may be taken to mean embryology, and phylogeny what we generally call evolution.

Further, the embryonic phenomena sometimes reproduce ancestral forms, and they are then called PALINGENETIC (from palin = again): sometimes they do not recall ancestral forms, but are later modifications due to adaptation, and they are then called CENOGENETIC (from kenos = new or foreign).

These terms are now widely used, but the reader of Haeckel must understand them thoroughly.

The first five chapters are an easy account of the history of embryology and evolution. The sixth and seventh give an equally clear account of the sexual elements and the process of conception. But some of the succeeding chapters must deal with embryonic processes so unfamiliar, and pursue them through so wide a range of animals in a brief space, that, in spite of the 200 illustrations, they will offer difficulty to many a reader. As our aim is to secure, not a superficial acquiescence in conclusions, but a fair comprehension of the truths of science, we have retained these chapters. However, I will give a brief and clear outline of the argument, so that the reader with little leisure may realise their value.

When the animal ovum (egg–cell) has been fertilised, it divides and subdivides until we have a cluster of cohering cells, externally not unlike a raspberry or mulberry. This is the morula (= mulberry) stage. The cluster becomes hollow, or filled with fluid in the centre, all the cells rising to the surface. This is the blastula (hollow ball) stage. One half of the cluster then bends or folds in upon the other, as one might do with a thin indiarubber ball, and we get a vase–shaped body with hollow interior (the first stomach, or "primitive gut"), an open mouth (the first or "primitive mouth"), and a wall composed of two layers of cells (two "germinal layers"). This is the gastrula (stomach) stage, and the process of its formation is called gastrulation. A glance at the illustration (Figure 1.29) will make this perfectly clear.

So much for the embryonic process in itself. The application to evolution has been a long and laborious task. Briefly, it was necessary to show that ALL the multicellular animals passed through these three stages, so that our biogenetic law would enable us to recognise them as reminiscences of ancestral forms. This is the work of Chapters 1.8 and 1.9. The difficulty can be realised in this way: As we reach the higher animals the ovum has to take up a large quantity of yelk, on which it may feed in developing. Think of the bird's "egg." The effect of this was to flatten the germ (the morula and blastula) from the first, and so give, at first sight, a totally different complexion to what it has in the lowest animals. When we pass the reptile and bird stage, the large yelk almost disappears (the germ now being supplied with blood by the mother), but the germ has been permanently altered in shape, and there are now a number of new embryonic processes (membranes, blood–vessel connections, etc.). Thus it was no light task to trace the identity of this process of gastrulation in all the animals. It has been done, however; and with this introduction the reader will be able to follow the proof. The conclusion is important. If all animals pass through the curious gastrula stage, it must be because they all had a common ancestor of that nature. To this conjectural ancestor (it lived before the period of fossilisation begins) Haeckel gives the name of the Gastraea, and in the second volume

11

we shall see a number of living animals of this type ("gastraeads").

The line of argument is the same in the next chapter. After laborious and careful research (though this stage is not generally admitted in the same sense as the previous one), a fourth common stage was discovered, and given the name of the Coelomula. The blastula had one layer of cells, the blastoderm (derma = skin): the gastrula two layers, the ectoderm ("outer skin") and entoderm ("inner skin"). Now a third layer (mesoderm = middle skin) is formed, by the growth inwards of two pouches or folds of the skin. The pouches blend together, and form a single cavity (the body cavity, or coelom), and its two walls are two fresh "germinal layers." Again, the identity of the process has to be proved in all the higher classes of animals, and when this is done we have another ancestral stage, the Coelomaea.

The remaining task is to build up the complex frame of the higher animals—always showing the identity of the process (on which the evolutionary argument depends) in enormously different conditions of embryonic life—out of the four "germinal layers." Chapter 1.9 prepares us for the work by giving us a very clear account of the essential structure of the back–boned (vertebrate) animal, and the probable common ancestor of all the vertebrates (a small fish of the lancelet type). Chapters 1.11 to 1.14 then carry out the construction step by step. The work is now simpler, in the sense that we leave all the invertebrate animals out of account; but there are so many organs to be fashioned out of the four simple layers that the reader must proceed carefully. In the second volume each of these organs will be dealt with separately, and the parallel will be worked out between its embryonic and its phylogenetic (evolutionary) development. The general reader may wait for this for a full understanding. But in the meantime the wonderful story of the construction of all our organs in the course of a few weeks (the human frame is perfectly formed, though less than two inches in length, by the twelfth week) from so simple a material is full of interest. It would be useless to attempt to summarise the process. The four chapters are themselves but a summary of it, and the eighty fine illustrations of the process will make it sufficiently clear. The last chapter carries the story on to the point where man at last parts company with the anthropoid ape, and gives a full account of the membranes or wrappers that enfold him in the womb, and the connection with the mother.

In conclusion, I would urge the reader to consult, at his free library perhaps, the complete edition of this work, when he has read the present abbreviated edition. Much of the text has had to be condensed in order to bring out the work at our popular price, and the beautiful plates of the complete edition have had to be omitted. The reader will find it an immense assistance if he can consult the library edition.

JOSEPH MCCABE.

Cricklewood, March, 1906.

***

HAECKEL'S CLASSIFICATION OF THE ANIMAL WORLD.

UNICELLULAR ANIMALS (PROTOZOA).

1. Unnucleated.

Bacteria.
Protamoebae.

Monera.

2. Nucleated.

2A. Rhizopoda.

Amoebina.
Radiolaria.

2B. Infusoria.

Flagellata.
Ciliata.

3. Cell–Colonies.

Catallacta.
Blastaeada.

MULTICELLULAR ANIMALS (METAZOA).

1. COELENTERIA, COELENTERATA, OR ZOOPHYTES.
Animals without body–cavity, blood or anus.

1A. Gastraeads.

Gastremaria.
Cyemaria.

1B. Sponges.

Protospongiae.
Metaspongiae.

1C. Cnidaria (Stinging Animals).

Hydrozoa.
Polyps.
Medusae.

1D. Platodes (Flat–Worms).

Platodaria.
Turbellaria.
Trematoda.
Cestoda.

2. COELOMARIA OR BILATERALS.
Animals with body–cavity and anus, and generally blood.

2A. Vermalia (Worm–Like).

Rotatoria.
Strongylaria.
Prosopygia.
Frontonia.

2B. Molluscs.

Cochlides.
Conchades.
Teuthodes.

2C. Articulates.

Annelida.
Crustacea.
Tracheata.

2D. Echinoderms.

Monorchonia.
Pentorchonia.

2E. Tunicates.

Copelata.
Ascidiae.
Thalidiae.

2F. Vertebrates.

2F.1. Acrania–Lancelet (Without Skull).

2F.2. Craniota (With Skull).

2F.2A. Cyclostomes. ("Round–Mouthed").

2F.2B. Fishes.

Selachii.
Ganoids.
Teleosts.
Dipneusts.

2F.2C. Amphibia.

2F.2D. Reptiles.

2F.2E. Birds.

2F.2F. Mammal.

Monotremes.

Marsupials.

Placentals:—
Rodents.
Edentates.
Ungulates.
Cetacea.
Sirenia.
Insectivora.
Cheiroptera.
Carnassia.
Primates.

(This classification is given for the purpose of explaining Haeckel's use of terms in this volume. The general reader should bear in mind that it differs very considerably from more recent schemes of classification. He should compare the scheme framed by Professor E. Ray Lankester.)

***

THE EVOLUTION OF MAN.

CHAPTER 1.1. THE FUNDAMENTAL LAW OF ORGANIC EVOLUTION.

The field of natural phenomena into which I would introduce my readers in the following chapters has a quite peculiar place in the broad realm of scientific inquiry. There is no object of investigation that touches man more closely, and the knowledge of which should be more acceptable to him, than his own frame. But among all the various branches of the natural history of mankind, or anthropology, the story of his development by natural means must excite the most lively interest. It gives us the key of the great world–riddles at which the human mind has been working for thousands of years. The problem of the nature of man, or the question of man's place in nature, and the cognate inquiries as to the past, the earliest history, the present situation, and the future of humanity—all these most important questions are directly and intimately connected with that branch of study which we call the science of the evolution of man, or, in one word, "Anthropogeny" (the genesis of man). Yet it is an astonishing fact that the science of the evolution of man does not even yet form part of the scheme of general education. In fact, educated people even in our day are for the most part quite ignorant of the important truths and remarkable phenomena which anthropogeny teaches us.

As an illustration of this curious state of things, it may be pointed out that most of what are considered to be "educated" people do not know that every human being is developed from an egg, or ovum, and that this egg is one simple cell, like any other plant or animal egg. They are equally ignorant that in the course of the development of this tiny, round egg–cell there is first formed a body that is totally different from the human frame, and has not the remotest resemblance to it. Most of them have never seen such a human embryo in the earlier period of its development, and do not know that it is quite indistinguishable from other animal embryos. At first the embryo is no more than a round cluster of cells, then it becomes a simple hollow sphere, the wall of which is composed of a layer of cells. Later it approaches very closely, at one period, to the anatomic structure of the lancelet, afterwards to that of a fish, and again to the typical build of the amphibia and mammals. As it continues to develop, a form appears which is like those we find at the lowest stage of mammal–life (such as the duck–bills), then a form that resembles the marsupials, and only at a late stage a form that has a resemblance to the ape; until at last the definite human form emerges and closes the series of transformations. These suggestive facts are, as I said, still almost unknown to the general public—so completely unknown that, if one casually mentions them, they are called in question or denied outright as fairy–tales. Everybody knows that the butterfly emerges from the pupa, and the pupa from a quite different thing called a larva, and the larva from the butterfly's egg. But few besides medical men are aware that MAN, in the course of his individual formation, passes through a series of transformations which are not less surprising and wonderful than the familiar metamorphoses of the butterfly.

The mere description of these remarkable changes through which man passes during his embryonic life should arouse considerable interest. But the mind will experience a far keener satisfaction when we trace these curious facts to their causes, and when we learn to behold in them natural phenomena which are of the highest importance throughout the whole field of human knowledge. They throw light first of all on the "natural history of creation," then on psychology, or "the science of the soul," and through this on the whole of philosophy. And as the general results of every branch of inquiry are summed up in philosophy, all the sciences come in turn to be touched and influenced more or less by the study of the evolution of man.

But when I say that I propose to present here the most important features of these phenomena and trace them to their causes, I take the term, and I interpret my task, in a very much wider sense than is usual. The lectures which have been delivered on this subject in the universities during the last half–century are almost exclusively adapted to medical men. Certainly, the medical man has the greatest interest in studying the origin of the human body, with which he is daily occupied. But I must not give here this special description of the embryonic processes such as it has hitherto been given, as most of my readers have not studied anatomy, and are not likely to be entrusted with the care of the adult organism. I must content myself with giving some parts of the subject only in general outline, and must not enter upon all the marvellous, but very intricate and not easily described, details that are found in the story of the development of the human frame. To understand these fully a knowledge of anatomy is needed. I will endeavour to be as plain as possible in dealing with this branch of science. Indeed, a sufficient general idea of the course of the embryonic development of man can be obtained without going too closely into the anatomic details. I trust we may be able to arouse the same interest in this delicate field of inquiry as has been excited already in other branches of science; though we shall meet more obstacles here than elsewhere.

The story of the evolution of man, as it has hitherto been expounded to medical students, has usually been confined to embryology—more correctly, ontogeny—or the science of the development of the individual human organism. But this is really only the first part of our task, the first half of the story of the evolution of man in that wider sense in which we understand it here. We must add as the second half—as another and not less important and interesting branch of the science of the evolution of the human stem—phylogeny: this may be described as the science of the evolution of the various animal forms from which the human organism has been developed in the course of countless ages. Everybody now knows of the great scientific activity that was occasioned by the publication of Darwin's Origin of Species in 1859. The chief direct consequence of this publication was to provoke a fresh inquiry into the origin of the human race, and this has proved beyond question our gradual evolution from the lower species. We give the name of "Phylogeny" to the science which describes this ascent of man from the lower ranks of the animal world. The chief source that it draws upon for facts is "Ontogeny," or embryology, the science of the development of the individual organism. Moreover, it derives a good deal of support from paleontology, or the science of fossil remains, and even more from comparative anatomy, or morphology.

These two branches of our science—on the one side ontogeny or embryology, and on the other phylogeny, or the science of race–evolution—are most vitally connected. The one cannot be understood without the other. It is only when the two branches fully co–operate and supplement each other that "Biogeny" (or the science of the genesis of life in the widest sense) attains to the rank of a philosophic science. The connection between them is not external and superficial, but profound, intrinsic, and causal. This is a discovery made by recent research, and it is most clearly and correctly expressed in the comprehensive law which I have called "the fundamental law of organic evolution," or "the fundamental law of biogeny." This general law, to which we shall find ourselves constantly recurring, and on the recognition of which depends one's whole insight into the story of evolution, may be briefly expressed in the phrase: "The history of the foetus is a recapitulation of the history of the race"; or, in other words, "Ontogeny is a recapitulation of phylogeny." It may be more fully stated as follows: The series of forms through which the individual organism passes during its development from the ovum to the complete bodily structure is a brief, condensed repetition of the long series of forms which the animal ancestors of the said organism, or the ancestral forms of the species, have passed through from the earliest period of organic life down to the present day.

The causal character of the relation which connects embryology with stem–history is due to the action of heredity and adaptation. When we have rightly understood these, and recognised their great importance in the formation of organisms, we can go a step further and say: Phylogenesis is the mechanical cause of ontogenesis.* (* The term "genesis," which occurs throughout, means, of course, "birth" or origin. From this we get: Biogeny = the origin of life (bios); Anthropogeny = the origin of man (anthropos); Ontogeny = the origin of the individual (on); Phylogeny = the origin of the species (phulon); and so on. In each case the term may refer to the process itself, or to the science describing the process.—Translator.) In other words, the development of the stem, or race, is, in accordance with the laws of heredity and adaptation, the cause of all the changes which appear in a condensed form in the evolution of the foetus.

The chain of manifold animal forms which represent the ancestry of each higher organism, or even of man, according to the theory of descent, always form a connected whole. We may designate this uninterrupted series of forms with the letters of the alphabet: A, B, C, D, E, etc., to Z. In apparent contradiction to what I have said, the story of the development of the individual, or the ontogeny of most organisms, only offers to the observer a part of these forms; so that the defective series of embryonic forms would run: A, B, D, F, H, K, M, etc.; or, in other cases, B, D, H, L, M, N, etc. Here, then, as a rule, several of the evolutionary forms of the original series have fallen out. Moreover, we often find—to continue with our illustration from the alphabet—one or other of the original letters of the ancestral series represented by corresponding letters from a different alphabet. Thus, instead of the Roman B and D, we often have the Greek Beta and Delta. In this case the text of the biogenetic law has been corrupted, just as it had been abbreviated in the preceding case. But, in spite of all this, the series of ancestral forms remains the same, and we are in a position to discover its original complexion.

In reality, there is always a certain parallel between the two evolutionary series. But it is obscured from the fact that in the embryonic succession much is wanting that certainly existed in the earlier ancestral succession. If the parallel of the two series were complete, and if this great fundamental law affirming the causal connection between ontogeny and phylogeny in the proper sense of the word were directly demonstrable, we should only have to determine, by means of the microscope and the dissecting knife, the series of forms through which the fertilised ovum passes in its development; we should then have before us a complete picture of the remarkable series of forms which our animal ancestors have successively assumed from the dawn of organic life down to the appearance of man. But such a repetition of the ancestral history by the individual in its embryonic life is very rarely complete. We do not often find our full alphabet. In most cases the correspondence is very imperfect, being greatly distorted and falsified by causes which we will consider later. We are thus, for the most part, unable to determine in detail, from the study of its embryology, all the different shapes which an organism's ancestors have assumed; we usually—and especially in the case of the human foetus—encounter many gaps. It is true that we can fill up most of these gaps satisfactorily with the help of comparative anatomy, but we cannot do so from direct embryological observation. Hence it is important that we find a large number of lower animal forms to be still represented in the course of man's embryonic development. In these cases we may draw our conclusions with the utmost security as to the nature of the ancestral form from the features of the form which the embryo momentarily assumes.

To give a few examples, we can infer from the fact that the human ovum is a simple cell that the first ancestor of our species was a tiny unicellular being, something like the amoeba. In the same way, we know, from the fact that the human foetus consists, at the first, of two simple cell–layers (the gastrula), that the gastraea, a form with two such layers, was certainly in the line of our ancestry. A later human embryonic form (the chordula) points just as clearly to a worm–like ancestor (the prochordonia), the nearest living relation of which is found among the actual ascidiae. To this succeeds a most important embryonic stage (acrania), in which our headless foetus presents, in the main, the structure of the lancelet. But we can only indirectly and approximately, with the aid of comparative anatomy and ontogeny, conjecture what lower forms enter into the chain of our ancestry between the gastraea and the chordula, and between this and the lancelet. In the course of the historical development many intermediate structures have gradually fallen out, which must certainly have been represented in our ancestry. But, in spite of these many, and sometimes very appreciable, gaps, there is no contradiction between the two successions. In fact, it is the chief purpose of this work to prove the real harmony and the original parallelism of the two. I hope to show, on a substantial basis of facts, that we can draw most important conclusions as to our genealogical tree from the actual and easily–demonstrable series of embryonic changes. We shall then be in a position to form a general idea of the wealth of animal forms which have figured in the direct line of our ancestry in the lengthy history of organic life.

In this evolutionary appreciation of the facts of embryology we must, of course, take particular care to distinguish sharply and clearly between the primitive, palingenetic (or ancestral) evolutionary processes and those due to cenogenesis.* (* Palingenesis = new birth, or re–incarnation (palin = again, genesis or genea = development); hence its application to the phenomena which are recapitulated by heredity from earlier ancestral forms. Cenogenesis = foreign or negligible development (kenos and genea); hence, those phenomena which come later in the story of life to disturb the inherited structure, by a fresh adaptation to environment.—Translator.) By palingenetic processes, or embryonic recapitulations, we understand all those phenomena in the development of the individual which are transmitted from one generation to another by heredity, and which, on that account, allow us to draw direct inferences as to corresponding structures in the development of the species. On the other hand, we give the name of cenogenetic processes, or embryonic variations, to all those phenomena in the foetal development that cannot be traced to inheritance from earlier species, but are due to the adaptation of the foetus, or the infant–form, to certain conditions of its embryonic development. These cenogenetic phenomena are foreign or later additions; they allow us to draw no direct inference whatever as to corresponding processes in our ancestral history, but rather hinder us from doing so.

This careful discrimination between the primary or palingenetic processes and the secondary or cenogenetic is of great importance for the purposes of the scientific history of a species, which has to draw conclusions from the available facts of embryology, comparative anatomy, and paleontology, as to the processes in the formation of the species in the remote past. It is of the same importance to the student of evolution as the careful distinction between genuine and spurious texts in the works of an ancient writer, or the purging of the real text from interpolations and alterations, is for the student of philology. It is true that this distinction has not yet been fully appreciated by many scientists. For my part, I regard it as the first condition for forming any just idea of the evolutionary process, and I believe that we must, in accordance with it, divide embryology into two sections—palingenesis, or the science of recapitulated forms; and cenogenesis, or the science of supervening structures.

To give at once a few examples from the science of man's origin in illustration of this important distinction, I may instance the following processes in the embryology of man, and of all the higher vertebrates, as palingenetic: the formation of the two primary germinal layers and of the primitive gut, the undivided structure of the dorsal nerve–tube, the appearance of a simple axial rod between the medullary tube and the gut, the temporary formation of the gill–clefts and arches, the primitive kidneys, and so on.* (* All these, and the following structures, will be fully described in later chapters.—Translator.) All these, and many other important structures, have clearly been transmitted by a steady heredity from the early ancestors of the mammal, and are, therefore, direct indications of the presence of similar structures in the history of the stem. On the other hand, this is certainly not the case with the following embryonic forms, which we must describe as cenogenetic processes: the formation of the yelk–sac, the allantois, the placenta, the amnion, the serolemma, and the chorion—or, generally speaking, the various foetal membranes and the corresponding changes in the blood vessels. Further instances are: the dual structure of the heart cavity, the temporary division of the plates of the primitive vertebrae and lateral plates, the secondary closing of the ventral and intestinal walls, the formation of the navel, and so on. All these and many other phenomena are certainly not traceable to similar structures in any earlier and completely–developed ancestral form, but have arisen simply by adaptation to the peculiar conditions of embryonic life (within the foetal membranes). In view of these facts, we may now give the following more precise expression to our chief law of biogeny: The evolution of the foetus (or ontogenesis) is a condensed and abbreviated recapitulation of the evolution of the stem (or phylogenesis); and this recapitulation is the more complete in proportion as the original development (or palingenesis) is preserved by a constant heredity; on the other hand, it becomes less complete in proportion as a varying adaptation to new conditions increases the disturbing factors in the development (or cenogenesis).

The cenogenetic alterations or distortions of the original palingenetic course of development take the form, as a rule, of a gradual displacement of the phenomena, which is slowly effected by adaptation to the changed conditions of embryonic existence during the course of thousands of years. This displacement may take place as regards either the position or the time of a phenomenon.

The great importance and strict regularity of the time–variations in embryology have been carefully studied recently by Ernest Mehnert, in his Biomechanik (Jena, 1898). He contends that our biogenetic law has not been impaired by the attacks of its opponents, and goes on to say: "Scarcely any piece of knowledge has contributed so much to the advance of embryology as this; its formulation is one of the most signal services to general biology. It was not until this law passed into the flesh and blood of investigators, and they had accustomed themselves to see a reminiscence of ancestral history in embryonic structures, that we witnessed the great progress which embryological research has made in the last two decades." The best proof of the correctness of this opinion is that now the most fruitful work is done in all branches of embryology with the aid of this biogenetic law, and that it enables students to attain every year thousands of brilliant results that they would never have reached without it.

It is only when one appreciates the cenogenetic processes in relation to the palingenetic, and when one takes careful account of the changes which the latter may suffer from the former, that the radical importance of the biogenetic law is recognised, and it is felt to be the most illuminating principle in the science of evolution. In this task of discrimination it is the silver thread in relation to which we can arrange all the phenomena of this realm of marvels—the "Ariadne thread," which alone enables us to find our way through this labyrinth of forms. Hence the brothers Sarasin, the zoologists, could say with perfect justice, in their study of the evolution of the Ichthyophis, that "the great biogenetic law is just as important for the zoologist in tracing long–extinct processes as spectrum analyses is for the astronomer."

Even at an earlier period, when a correct acquaintance with the evolution of the human and animal frame was only just being obtained—and that is scarcely eighty years ago!—the greatest astonishment was felt at the remarkable similarity observed between the embryonic forms, or stages of foetal development, in very different animals; attention was called even then to their close resemblance to certain fully–developed animal forms belonging to some of the lower groups. The older scientists (Oken, Treviranus, and others) knew perfectly well that these lower forms in a sense illustrated and fixed, in the hierarchy of the animal world, a temporary stage in the evolution of higher forms. The famous anatomist Meckel spoke in 1821 of a "similarity between the development of the embryo and the series of animals." Baer raised the question in 1828 how far, within the vertebrate type, the embryonic forms of the higher animals assume the permanent shapes of members of lower groups. But it was impossible fully to understand and appreciate this remarkable resemblance at that time. We owe our capacity to do this to the theory of descent; it is this that puts in their true light the action of heredity on the one hand and adaptation on the other. It explains to us the vital importance of their constant reciprocal action in the production of organic forms. Darwin was the first to teach us the great part that was played in this by the ceaseless struggle for existence between living things, and to show how, under the influence of this (by natural selection), new species were produced and maintained solely by the interaction of heredity and adaptation. It was thus Darwinism that first opened our eyes to a true comprehension of the supremely important relations between the two parts of the science of organic evolution—Ontogeny and Phylogeny.

Heredity and adaptation are, in fact, the two constructive physiological functions of living things; unless we understand these properly we can make no headway in the study of evolution. Hence, until the time of Darwin no one had a clear idea of the real nature and causes of embryonic development. It was impossible to explain the curious series of forms through which the human embryo passed; it was quite unintelligible why this strange succession of animal–like forms appeared in the series at all. It had previously been generally assumed that the man was found complete in all his parts in the ovum, and that the development consisted only in an unfolding of the various parts, a simple process of growth. This is by no means the case. On the contrary, the whole process of the development of the individual presents to the observer a connected succession of different animal–forms; and these forms display a great variety of external and internal structure. But WHY each individual human being should pass through this series of forms in the course of his embryonic development it was quite impossible to say until Lamarck and Darwin established the theory of descent. Through this theory we have at last detected the real causes, the efficient causes, of the individual development; we have learned that these mechanical causes suffice of themselves to effect the formation of the organism, and that there is no need of the final causes which were formerly assumed. It is true that in the academic philosophies of our time these final causes still figure very prominently; in the new philosophy of nature we can entirely replace them by efficient causes. We shall see, in the course of our inquiry, how the most wonderful and hitherto insoluble enigmas in the human and animal frame have proved amenable to a mechanical explanation, by causes acting without prevision, through Darwin's reform of the science of evolution. We have everywhere been able to substitute unconscious causes, acting from necessity, for conscious, purposive causes.* (* The monistic or mechanical philosophy of nature holds that only unconscious, necessary, efficient causes are at work in the whole field of nature, in organic life as well as in inorganic changes. On the other hand, the dualist or vitalist philosophy of nature affirms that unconscious forces are only at work in the inorganic world, and that we find conscious, purposive, or final causes in organic nature.)

If the new science of evolution had done no more than this, every thoughtful man would have to admit that it had accomplished an immense advance in knowledge. It means that in the whole of philosophy that tendency which we call monistic, in opposition to the dualistic, which has hitherto prevailed, must be accepted.* (* Monism is neither purely materialistic nor purely spiritualistic, but a reconciliation of these two principles, since it regards the whole of nature as one, and sees only efficient causes at work in it. Dualism, on the contrary, holds that nature and spirit, matter and force, the world and God, inorganic and organic nature, are separate and independent existences. Cf. The Riddle of the Universe chapter 12.) At this point the science of human evolution has a direct and profound bearing on the foundations of philosophy. Modern anthropology has, by its astounding discoveries during the second half of the nineteenth century, compelled us to take a completely monistic view of life. Our bodily structure and its life, our embryonic development and our evolution as a species, teach us that the same laws of nature rule in the life of man as in the rest of the universe. For this reason, if for no others, it is desirable, nay, indispensable, that every man who wishes to form a serious and philosophic view of life, and, above all, the expert philosopher, should acquaint himself with the chief facts of this branch of science.

The facts of embryology have so great and obvious a significance in this connection that even in recent years dualist and teleological philosophers have tried to rid themselves of them by simply denying them. This was done, for instance, as regards the fact that man is developed from an egg, and that this egg or ovum is a simple cell, as in the case of other animals. When I had explained this pregnant fact and its significance in my History of Creation, it was described in many of the theological journals as a dishonest invention of my own. The fact that the embryos of man and the dog are, at a certain stage of their development, almost indistinguishable was also denied. When we examine the human embryo in the third or fourth week of its development, we find it to be quite different in shape and structure from the full-grown human being, but almost identical with that of the ape, the dog, the rabbit, and other mammals, at the same stage of ontogeny. We find a bean-shaped body of very simple construction, with a tail below and a pair of fins at the sides, something like those of a fish, but very different from the limbs of man and the mammals. Nearly the whole front half of the body is taken up by a shapeless head without face, at the sides of which we find gill-clefts and arches as in the fish. At this stage of its development the human embryo does not differ in any essential detail from that of the ape, dog, horse, ox, etc., at a corresponding period. This important fact can easily be verified at any moment by a comparison of the embryos of man, the dog, rabbit, etc. Nevertheless, the theologians and dualist philosophers pronounced it to be a materialistic invention; even scientists, to whom the facts should be known, have sought to deny them.

There could not be a clearer proof of the profound importance of these embryological facts in favour of the monistic philosophy than is afforded by these efforts of its opponents to get rid of them by silence or denial. The truth is that these facts are most inconvenient for them, and are quite irreconcilable with their views. We must be all the more pressing on our side to put them in their proper light. I fully agree with Huxley when he says, in his "Man's Place in Nature": "Though these facts are ignored by several well–known popular leaders, they are easy to prove, and are accepted by all scientific men; on the other hand, their importance is so great that those who have once mastered them will, in my opinion, find few other biological discoveries to astonish them."

We shall make it our chief task to study the evolution of man's bodily frame and its various organs in their external form and internal structures. But I may observe at once that this is accompanied step by step with a study of the evolution of their functions. These two branches of inquiry are inseparably united in the whole of anthropology, just as in zoology (of which the former is only a section) or general biology. Everywhere the peculiar form of the organism and its structures, internal and external, is directly related to the special physiological functions which the organism or organ has to execute. This intimate connection of structure and function, or of the instrument and the work done by it, is seen in the science of evolution and all its parts. Hence the story of the evolution of structures, which is our immediate concern, is also the history of the development of functions; and this holds good of the human organism as of any other.

At the same time, I must admit that our knowledge of the evolution of functions is very far from being as complete as our acquaintance with the evolution of structures. One might say, in fact, that the whole science of evolution has almost confined itself to the study of structures; the evolution of FUNCTIONS hardly exists even in name. That is the fault of the physiologists, who have as yet concerned themselves very little about evolution. It is only in recent times that physiologists like W. Engelmann, W. Preyer, M. Verworn, and a few others, have attacked the evolution of functions.

It will be the task of some future physiologist to engage in the study of the evolution of functions with the same zeal and success as has been done for the evolution of structures in morphogeny (the science of the genesis of forms). Let me illustrate the close connection of the two by a couple of examples. The heart in the human embryo has at first a very simple construction, such as we find in permanent form among the ascidiae and other low organisms; with this is associated a very simple system of circulation of the blood. Now, when we find that with the full–grown heart there comes a totally different and much more intricate circulation, our inquiry into the development of the heart becomes at once, not only an anatomical, but also a physiological, study. Thus it is clear that the ontogeny of the heart can only be understood in the light of its phylogeny (or development in the past), both as regards function and structure. The same holds true of all the other organs and their functions. For instance, the science of the evolution of the alimentary canal, the lungs, or the sexual organs, gives us at the same time, through the exact comparative investigation of structure–development, most important information with regard to the evolution of the functions of these organs.

This significant connection is very clearly seen in the evolution of the nervous system. This system is in the economy of the human body the medium of sensation, will, and even thought, the highest of the psychic functions; in a word, of all the various functions which constitute the proper object of psychology. Modern anatomy and physiology have proved that these psychic functions are immediately dependent on the fine structure and the composition of the central nervous system, or the internal texture of the brain and spinal cord. In these we find the elaborate cell–machinery, of which the psychic or soul–life is the physiological function. It is so intricate that most men still look upon the mind as something supernatural that cannot be explained on mechanical principles.

But embryological research into the gradual appearance and the formation of this important system of organs yields the most astounding and significant results. The first sketch of a central nervous system in the human embryo presents the same very simple type as in the other vertebrates. A spinal tube is formed in the external skin of the back, and from this first comes a simple spinal cord without brain, such as we find to be the permanent psychic organ in the lowest type of vertebrate, the amphioxus. Not until a later stage is a brain formed at the anterior end of this cord, and then it is a brain of the most rudimentary kind, such as we find permanently among the lower fishes. This simple brain develops step by step, successively assuming forms which correspond to those of the amphibia, the reptiles, the duck–bills, and the lemurs. Only in the last stage does it reach the highly organised form which distinguishes the apes from the other vertebrates, and which attains its full development in man.

Comparative physiology discovers a precisely similar growth. The function of the brain, the psychic activity, rises step by step with the advancing development of its structure.

Thus we are enabled, by this story of the evolution of the nervous system, to understand at length THE NATURAL DEVELOPMENT OF THE HUMAN MIND and its gradual unfolding. It is only with the aid of embryology that we can grasp how these highest and most striking faculties of the animal organism have been historically evolved. In other words, a knowledge of the evolution of the spinal cord and brain in the human embryo leads us directly to a comprehension of the historic development (or phylogeny) of the human mind, that highest of all faculties, which we regard as something so marvellous and supernatural in the adult man. This is certainly one of the greatest and most pregnant results of evolutionary science. Happily our embryological knowledge of man's central nervous system is now so adequate, and agrees so thoroughly with the complementary results of comparative anatomy and physiology, that we are thus enabled to obtain a clear insight into one of the highest problems of philosophy, the phylogeny of the soul, or the ancestral history of the mind of man. Our chief support in this comes from the embryological study of it, or the ontogeny of the soul. This important section of psychology owes its origin especially to W. Preyer, in his interesting works, such as The Mind of the Child. The Biography of a Baby (1900), of Milicent Washburn Shinn, also deserves mention. [See also Preyer's Mental Development in the Child (translation), and Sully's Studies of Childhood and Children's Ways.]

In this way we follow the only path along which we may hope to reach the solution of this difficult problem.

Thirty–six years have now elapsed since, in my General Morphology, I established phylogeny as an independent science and showed its intimate causal connection with ontogeny; thirty years have passed since I gave in my gastraea–theory the proof of the justice of this, and completed it with the theory of germinal layers. When we look back on this period we may ask, What has been accomplished during it by the fundamental law of biogeny? If we are impartial, we must reply that it has proved its fertility in hundreds of sound results, and that by its aid we have acquired a vast fund of knowledge which we should never have obtained without it.

There has been no dearth of attacks—often violent attacks—on my conception of an intimate causal connection between ontogenesis and phylogenesis; but no other satisfactory explanation of these important phenomena has yet been offered to us. I say this especially with regard to Wilhelm His's theory of a "mechanical evolution," which questions the truth of phylogeny generally, and would explain the complicated embryonic processes without going beyond by simple physical changes—such as the bending and folding of leaves by electricity, the origin of cavities through unequal strain of the tissues, the formation of processes by uneven growth, and so on. But the fact is that these embryological phenomena themselves demand explanation in turn, and this can only be found, as a rule, in the corresponding changes in the long ancestral series, or in the physiological functions of heredity and adaptation.

CHAPTER 1.2. THE OLDER EMBRYOLOGY.

It is in many ways useful, on entering upon the study of any science, to cast a glance at its historical development. The saying that "everything is best understood in its growth" has a distinct application to science. While we follow its gradual development we get a clearer insight into its aims and objects. Moreover, we shall see that the present condition of the science of human evolution, with all its characteristics, can only be rightly understood when we examine its historical growth. This task will, however, not detain us long. The study of man's evolution is one of the latest branches of natural science, whether you consider the embryological or the phylogenetic section of it.

Apart from the few germs of our science which we find in classical antiquity, and which we shall notice presently, we may say that it takes its definite rise, as a science, in the year 1759, when one of the greatest German scientists, Caspar Friedrich Wolff, published his Theoria generationis. That was the foundation–stone of the science of animal embryology. It was not until fifty years later, in 1809, that Jean Lamarck published his Philosophie Zoologique—the first effort to provide a base for the theory of evolution; and it was another half–century before Darwin's work appeared (in 1859), which we may regard as the first scientific attainment of this aim. But before we go further into this solid establishment of evolution, we must cast a brief glance at that famous philosopher and scientist of antiquity, who stood alone in this, as in many other branches of science, for more than 2000 years: the "father of Natural History," Aristotle.

The extant scientific works of Aristotle deal with many different sides of biological research; the most comprehensive of them is his famous History of Animals. But not less interesting is the smaller work, On the Generation of Animals (Peri zoon geneseos). This work treats especially of embryonic development, and it is of great interest as being the earliest of its kind and the only one that has come down to us in any completeness from classical antiquity.

Aristotle studied embryological questions in various classes of animals, and among the lower groups he learned many most remarkable facts which we only rediscovered between 1830 and 1860. It is certain, for instance, that he was acquainted with the very peculiar mode of propagation of the cuttlefishes, or cephalopods, in which a yelk−sac hangs out of the mouth of the foetus. He knew, also, that embryos come from the eggs of the bee even when they have not been fertilised. This "parthenogenesis" (or virgin−birth) of the bees has only been established in our time by the distinguished zoologist of Munich, Siebold. He discovered that male bees come from the unfertilised, and female bees only from the fertilised, eggs. Aristotle further states that some kinds of fishes (of the genus serranus) are hermaphrodites, each individual having both male and female organs and being able to fertilise itself; this, also, has been recently confirmed. He knew that the embryo of many fishes of the shark family is attached to the mother's body by a sort of placenta, or nutritive organ very rich in blood; apart from these, such an arrangement is only found among the higher mammals and man. This placenta of the shark was looked upon as legendary for a long time, until Johannes Muller proved it to be a fact in 1839. Thus a number of remarkable discoveries were found in Aristotle's embryological work, proving a very good acquaintance of the great scientist—possibly helped by his predecessors—with the facts of ontogeny, and a great advance upon succeeding generations in this respect.

In the case of most of these discoveries he did not merely describe the fact, but added a number of observations on its significance. Some of these theoretical remarks are of particular interest, because they show a correct appreciation of the nature of the embryonic processes. He conceives the development of the individual as a new formation, in the course of which the various parts of the body take shape successively. When the human or animal frame is developed in the mother's body, or separately in an egg, the heart—which he regards as the starting−point and centre of the organism—must appear first. Once the heart is formed the other organs arise, the internal ones before the external, the upper (those above the diaphragm) before the lower (or those beneath the diaphragm). The brain is formed at an early stage, and the eyes grow out of it. These observations are quite correct. And, if we try to form some idea from these data of

Aristotle's general conception of the embryonic process, we find a dim prevision of the theory which Wolff showed 2000 years afterwards to be the correct view. It is significant, for instance, that Aristotle denied the eternity of the individual in any respect. He said that the species or genus, the group of similar individuals, might be eternal, but the individual itself is temporary. It comes into being in the act of procreation, and passes away at death.

During the 2000 years after Aristotle no progress whatever was made in general zoology, or in embryology in particular. People were content to read, copy, translate, and comment on Aristotle. Scarcely a single independent effort at research was made in the whole of the period. During the Middle Ages the spread of strong religious beliefs put formidable obstacles in the way of independent scientific investigation. There was no question of resuming the advance of biology. Even when human anatomy began to stir itself once more in the sixteenth century, and independent research was resumed into the structure of the developed body, anatomists did not dare to extend their inquiries to the unformed body, the embryo, and its development. There were many reasons for the prevailing horror of such studies. It is natural enough, when we remember that a Bull of Boniface VIII excommunicated every man who ventured to dissect a human corpse. If the dissection of a developed body were a crime to be thus punished, how much more dreadful must it have seemed to deal with the embryonic body still enclosed in the womb, which the Creator himself had decently veiled from the curiosity of the scientist! The Christian Church, then putting many thousands to death for unbelief, had a shrewd presentiment of the menace that science contained against its authority. It was powerful enough to see that its rival did not grow too quickly.

It was not until the Reformation broke the power of the Church, and a refreshing breath of the spirit dissolved the icy chains that bound science, that anatomy and embryology, and all the other branches of research, could begin to advance once more. However, embryology lagged far behind anatomy. The first works on embryology appear at the beginning of the sixteenth century. The Italian anatomist, Fabricius ab Aquapendente, a professor at Padua, opened the advance. In his two books (De formato foetu, 1600, and De formatione foetus, 1604) he published the older illustrations and descriptions of the embryos of man and other mammals, and of the hen. Similar imperfect illustrations were given by Spigelius (De formato foetu, 1631), and by Needham (1667) and his more famous compatriot, Harvey (1652), who discovered the circulation of the blood in the animal body and formulated the important principle, Omne vivum ex vivo (all life comes from pre—existing life). The Dutch scientist, Swammerdam, published in his Bible of Nature the earliest observations on the embryology of the frog and the division of its egg—yelk. But the most important embryological studies in the sixteenth century were those of the famous Italian, Marcello Malpighi, of Bologna, who led the way both in zoology and botany. His treatises, De formatione pulli and De ovo incubato (1687), contain the first consistent description of the development of the chick in the fertilised egg.

Here I ought to say a word about the important part played by the chick in the growth of our science. The development of the chick, like that of the young of all other birds, agrees in all its main features with that of the other chief vertebrates, and even of man. The three highest classes of vertebrates—mammals, birds, and reptiles (lizards, serpents, tortoises, etc.)—have from the beginning of their embryonic development so striking a resemblance in all the chief points of structure, and especially in their first forms, that for a long time it is impossible to distinguish between them. We have known now for some time that we need only examine the embryo of a bird, which is the easiest to get at, in order to learn the typical mode of development of a mammal (and therefore of man). As soon as scientists began to study the human embryo, or the mammal—embryo generally, in its earlier stages about the middle and end of the seventeenth century, this important fact was very quickly discovered. It is both theoretically and practically of great value. As regards the THEORY of evolution, we can draw the most weighty inferences from this similarity between the embryos of widely different classes of animals. But for the practical purposes of embryological research the discovery is invaluable, because we can fill up the gaps in our imperfect knowledge of the embryology of the mammals from the more thoroughly studied embryology of the bird. Hens' eggs are easily to be had in any quantity, and the development of the chick may be followed step by step in artificial incubation. The development of the mammal is much more difficult to follow, because here the embryo is not detached and enclosed in a large egg, but the tiny ovum remains in the womb until the growth is completed. Hence, it is very difficult to keep up sustained observation of the various stages in any great extent, quite apart from such extrinsic considerations as the cost, the technical difficulties, and many other obstacles which we encounter when we would make an extensive study of the fertilised mammal. The chicken has, therefore, always been the chief object of study in this connection. The excellent incubators we now have enable us to observe it in any quantity and at any stage of development, and so follow the whole course of its formation step by step.

By the end of the seventeenth century Malpighi had advanced as far as it was possible to do with the imperfect microscope of his time in the embryological study of the chick. Further progress was arrested until the instrument and the technical methods should be improved. The vertebrate embryos are so small and delicate in their earlier stages that you cannot go very far into the study of them without a good microscope and other technical aid. But this substantial improvement of the microscope and the other apparatus did not take place until the beginning of the nineteenth century.

Embryology made scarcely any advance in the first half of the eighteenth century, when the systematic natural history of plants and animals received so great an impulse through the publication of Linne's famous Systema Naturae. Not until 1759 did the genius arise who was to give it an entirely new character, Caspar Friedrich Wolff. Until then embryology had been occupied almost exclusively in unfortunate and misleading efforts to build up theories on the imperfect empirical material then available.

The theory which then prevailed, and remained in favour throughout nearly the whole of the eighteenth century, was commonly called at that time "the evolution theory"; it is better to describe it as "the preformation theory."* (* This theory is usually known as the "evolution theory" in Germany, in contradistinction to the "epigenesis theory." But as it is the latter that is called the "evolution theory" in England, France, and Italy, and "evolution" and "epigenesis" are taken to be synonymous, it seems better to call the first the "pre−formation theory.") Its chief point is this: There is no new formation of structures in the embryonic development of any organism, animal or plant, or even of man; there is only a growth, or unfolding, of parts which have been constructed or pre−formed from all eternity, though on a very small scale and closely packed together. Hence, every living germ contains all the organs and parts of the body, in the form and arrangement they will present later, already within it, and thus the whole embryological process is merely an evolution in the literal sense of the word, or an unfolding, of parts that were pre−formed and folded up in it. So, for instance, we find in the hen's egg not merely a simple cell, that divides and subdivides and forms germinal layers, and at last, after all kinds of variation and cleavage and reconstruction, brings forth the body of the chick; but there is in every egg from the first a complete chicken, with all its parts made and neatly packed. These parts are so small or so transparent that the microscope cannot detect them. In the hatching, these parts merely grow larger, and spread out in the normal way.

When this theory is consistently developed it becomes a "scatulation theory."* (* "Packing theory" would be the literal translation. Scatula is the Latin for a case or box.—Translator.) According to its teaching, there was made in the beginning one couple or one individual of each species of animal or plant; but this one individual contained the germs of all the other individuals of the same species who should ever come to life. As the age of the earth was generally believed at that time to be fixed by the Bible at 5000 or 6000 years, it seemed possible to calculate how many individuals of each species had lived in the period, and so had been packed inside the first being that was created. The theory was consistently extended to man, and it was affirmed that our common parent Eve had had stored in her ovary the germs of all the children of men.

The theory at first took the form of a belief that it was the FEMALES who were thus encased in the first being. One couple of each species was created, but the female contained in her ovary all the future individuals of the species, of either sex. However, this had to be altered when the Dutch microscopist, Leeuwenhoek, discovered the male spermatozoa in 1690, and showed that an immense number of these extremely fine and mobile thread–like beings exist in the male sperm (this will be explained in Chapter 2.7). This astonishing discovery was further advanced when it was proved that these living bodies, swimming about in the seminal fluid, were real animalcules, and, in fact, were the pre–formed germs of the future generation. When the male and female procreative elements came together at conception, these thread–like spermatozoa ("seed–animals") were supposed to penetrate into the fertile body of the ovum and begin to develop there, as the plant seed does in the fruitful earth. Hence, every spermatozoon was regarded as a homunculus, a tiny complete man; all the parts were believed to be pre–formed in it, and merely grew larger when it reached its proper medium in the female ovum. This theory, also, was consistently developed in the sense that in each of these thread–like bodies the whole of its posterity was supposed to be present in the minutest form. Adam's sexual glands were thought to have contained the germs of the whole of humanity.

This "theory of male scatulation" found itself at once in keen opposition to the prevailing "female" theory. The two rival theories at once opened a very lively campaign, and the physiologists of the eighteenth century were divided into two great camps—the Animalculists and the Ovulists—which fought vigorously. The animalculists held that the spermatozoa were the true germs, and appealed to the lively movements and the structure of these bodies. The opposing party of the Ovulists, who clung to the older "evolution theory," affirmed that the ovum is the real germ, and that the spermatozoa merely stimulate it at conception to begin its growth; all the future generations are stored in the ovum. This view was held by the great majority of the biologists of the eighteenth century, in spite of the fact that Wolff proved it in 1759 to be without foundation. It owed its prestige chiefly to the circumstance that the most weighty authorities in the biology and philosophy of the day decided in favour of it, especially Haller, Bonnet, and Leibnitz.

Albrecht Haller, professor at Gottingen, who is often called the father of physiology, was a man of wide and varied learning, but he does not occupy a very high position in regard to insight into natural phenomena. He made a vigorous defence of the "evolutionary theory" in his famous work, Elementa physiologiae, affirming: "There is no such thing as formation (nulla est epigenesis). No part of the animal frame is made before another; all were made together." He thus denied that there was any evolution in the proper sense of the word, and even went so far as to say that the beard existed in the new-born child and the antlers in the hornless fawn; all the parts were there in advance, and were merely hidden from the eye of man for the time being. Haller even calculated the number of human beings that God must have created on the sixth day and stored away in Eve's ovary. He put the number at 200,000 millions, assuming the age of the world to be 6000 years, the average age of a human being to be thirty years, and the population of the world at that time to be 1000 millions. And the famous Haller maintained all this nonsense, in spite of its ridiculous consequences, even after Wolff had discovered the real course of embryonic development and established it by direct observation!

Among the philosophers of the time the distinguished Leibnitz was the chief defender of the "preformation theory," and by his authority and literary prestige won many adherents to it. Supported by his system of monads, according to which body and soul are united in inseparable association and by their union form the individual, or the "monad," Leibnitz consistently extended the "scatulation theory" to the soul, and held that this was no more evolved than the body. He says, for instance, in his Theodicee: "I mean that these souls, which one day are to be the souls of men, are present in the seed, like those of other species; in such wise that they existed in our ancestors as far back as Adam, or from the beginning of the world, in the forms of organised bodies."

The theory seemed to receive considerable support from the observations of one of its most zealous supporters, Bonnet. In 1745 he discovered, in the plant-louse, a case of parthenogenesis, or virgin-birth, an interesting form of reproduction that has lately been found by Siebold and others among various classes of the articulata, especially crustacea and insects. Among these and other animals of certain lower species the female may reproduce for several generations without having been fertilised by the male. These ova that do not need fertilisation are called "false ova," pseudova or spores. Bonnet saw that a female plant–louse, which he had kept in cloistral isolation, and rigidly removed from contact with males, had on the eleventh day (after forming a new skin for the fourth time) a living daughter, and during the next twenty days ninety–four other daughters; and that all of them went on to reproduce in the same way without any contact with males. It seemed as if this furnished an irrefutable proof of the truth of the scatulation theory, as it was held by the Ovulists; it is not surprising to find that the theory then secured general acceptance.

This was the condition of things when suddenly, in 1759, Caspar Friedrich Wolff appeared, and dealt a fatal blow at the whole preformation theory with his new theory of epigenesis. Wolff, the son of a Berlin tailor, was born in 1733, and went through his scientific and medical studies, first at Berlin under the famous anatomist Meckel, and afterwards at Halle. Here he secured his doctorate in his twenty–sixth year, and in his academic dissertation (November 28th, 1759), the Theoria generationis, expounded the new theory of a real development on a basis of epigenesis. This treatise is, in spite of its smallness and its obscure phraseology, one of the most valuable in the whole range of biological literature. It is equally distinguished for the mass of new and careful observations it contains, and the far–reaching and pregnant ideas which the author everywhere extracts from his observations and builds into a luminous and accurate theory of generation. Nevertheless, it met with no success at the time. Although scientific studies were then assiduously cultivated owing to the impulse given by Linne—although botanists and zoologists were no longer counted by dozens, but by hundreds, hardly any notice was taken of Wolff's theory. Even when he established the truth of epigenesis by the most rigorous observations, and demolished the airy structure of the preformation theory, the "exact" scientist Haller proved one of the most strenuous supporters of the old theory, and rejected Wolff's correct view with a dictatorial "There is no such thing as evolution." He even went on to say that religion was menaced by the new theory! It is not surprising that the whole of the physiologists of the second half of the eighteenth century submitted to the ruling of this physiological pontiff, and attacked the theory of epigenesis as a dangerous innovation. It was not until more than fifty years afterwards that Wolff's work was appreciated. Only when Meckel translated into German in 1812 another valuable work of Wolff's on The Formation of the Alimentary Canal (written in 1768), and called attention to its great importance, did people begin to think of him once more;

yet this obscure writer had evinced a profounder insight into the nature of the living organism than any other scientist of the eighteenth century.

Wolff's idea led to an appreciable advance over the whole field of biology. There is such a vast number of new and important observations and pregnant thoughts in his writings that we have only gradually learned to appreciate them rightly in the course of the nineteenth century. He opened up the true path for research in many directions. In the first place, his theory of epigenesis gave us our first real insight into the nature of embryonic development. He showed convincingly that the development of every organism consists of a series of NEW FORMATIONS, and that there is no trace whatever of the complete form either in the ovum or the spermatozoon. On the contrary, these are quite simple bodies, with a very different purport. The embryo which is developed from them is also quite different, in its internal arrangement and outer configuration, from the complete organism. There is no trace whatever of preformation or in–folding of organs. To–day we can scarcely call epigenesis a THEORY, because we are convinced it is a fact, and can demonstrate it at any moment with the aid of the microscope.

Wolff furnished the conclusive empirical proof of his theory in his classic dissertation on The Formation of the Alimentary Canal (1768). In its complete state the alimentary canal of the hen is a long and complex tube, with which the lungs, liver, salivary glands, and many other small glands, are connected. Wolff showed that in the early stages of the embryonic chick there is no trace whatever of this complicated tube with all its dependencies, but instead of it only a flat, leaf–shaped body; that, in fact, the whole embryo has at first the appearance of a flat, oval–shaped leaf. When we remember how difficult the exact observation of so fine and delicate a structure as the early leaf–shaped body of the chick must have been with the poor microscopes then in use, we must admire the rare faculty for observation which enabled Wolff to make the most important discoveries in this most difficult part of embryology. By this laborious research he reached the correct opinion that the embryonic body of all the higher animals, such as the birds, is for some time merely a flat, thin, leaf–shaped disk—consisting at first of one layer, but afterwards of several. The lowest of these layers is the alimentary canal, and Wolff followed its development from its commencement to its completion. He showed how this leaf–shaped structure first turns into a groove, then the margins of this groove fold together and form a closed canal, and at length the two external openings of the tube (the mouth and anus) appear.

Moreover, the important fact that the other systems of organs are developed in the same way, from tubes formed out of simple layers, did not escape Wolff. The nerveless system, muscular system, and vascular (blood–vessel) system, with all the organs appertaining thereto, are, like the alimentary system, developed out of simple leaf–shaped structures. Hence, Wolff came to the view by 1768 which Pander developed in the Theory of Germinal Layers fifty years afterwards. His principles are not literally correct; but he comes as near to the truth in them as was possible at that time, and could be expected of him.

Our admiration of this gifted genius increases when we find that he was also the precursor of Goethe in regard to the metamorphosis of plants and of the famous cellular theory. Wolff had, as Huxley showed, a clear presentiment of this cardinal theory, since he recognised small microscopic globules as the elementary parts out of which the germinal layers arose.

Finally, I must invite special attention to the MECHANICAL character of the profound philosophic reflections which Wolff always added to his remarkable observations. He was a great monistic philosopher, in the best meaning of the word. It is unfortunate that his philosophic discoveries were ignored as completely as his observations for more than half a century. We must be all the more careful to emphasise the fact of their clear monistic tendency.

CHAPTER 1.3. MODERN EMBRYOLOGY.

We may distinguish three chief periods in the growth of our science of human embryology. The first has been considered in the preceding chapter; it embraces the whole of the preparatory period of research, and extends from Aristotle to Caspar Friedrich Wolff, or to the year 1759, in which the epoch–making Theoria generationis was published. The second period, with which we have now to deal, lasts about a century—that is to say, until the appearance of Darwin's Origin of Species, which brought about a change in the very foundations of biology, and, in particular, of embryology. The third period begins with Darwin. When we say that the second period lasted a full century, we must remember that Wolff's work had remained almost unnoticed during half the time—namely, until the year 1812. During the whole of these fifty–three years not a single book that appeared followed up the path that Wolff had opened, or extended his theory of embryonic development. We merely find his views—perfectly correct views, based on extensive observations of fact—mentioned here and there as erroneous; their opponents, who adhered to the dominant theory of preformation, did not even deign to reply to them. This unjust treatment was chiefly due to the extraordinary authority of Albrecht von Haller; it is one of the most astonishing instances of a great authority, as such, preventing for a long time the recognition of established facts.

The general ignorance of Wolff's work was so great that at the beginning of the nineteenth century two scientists of Jena, Oken (1806) and Kieser (1810), began independent research into the development of the alimentary canal of the chick, and hit upon the right clue to the embryonic puzzle, without knowing a word about Wolff's important treatise on the same subject. They were treading in his very footsteps without suspecting it. This can be easily proved from the fact that they did not travel as far as Wolff. It was not until Meckel translated into German Wolff's book on the alimentary system, and pointed out its great importance, that the eyes of anatomists and physiologists were suddenly opened. At once a number of biologists instituted fresh embryological inquiries, and began to confirm Wolff's theory of epigenesis.

This resuscitation of embryology and development of the epigenesis–theory was chiefly connected with the university of Wurtzburg. One of the professors there at that time was Dollinger, an eminent biologist, and father of the famous Catholic historian who later distinguished himself by his opposition to the new dogma of papal infallibility. Dollinger was both a profound thinker and an accurate observer. He took the keenest interest in embryology, and worked at it a good deal. However, he is not himself responsible for any important result in this field. In 1816 a young medical doctor, whom we may at once designate as Wolff's chief successor, Karl Ernst von Baer, came to Wurtzburg. Baer's conversations with Dollinger on embryology led to a fresh series of most extensive investigations. Dollinger had expressed a wish that some young scientist should begin again under his guidance an independent inquiry into the development of the chick during the hatching of the egg. As neither he nor Baer had money enough to pay for an incubator and the proper control of the experiments, and for a competent artist to illustrate the various stages observed, the lead of the enterprise was given to Christian Pander, a wealthy friend of Baer's who had been induced by Baer to come to Wurtzburg. An able engraver, Dalton, was engaged to do the copper–plates. In a short time the embryology of the chick, in which Baer was taking the greatest indirect interest, was so far advanced that

Pander was able to sketch the main features of it on the ground of Wolff's theory in the dissertation he published in 1817. He clearly enunciated the theory of germinal layers which Wolff had anticipated, and established the truth of Wolff's idea of a development of the complicated systems of organs out of simple leaf—shaped primitive structures. According to Pander, the leaf—shaped object in the hen's egg divides, before the incubation has proceeded twelve hours, into two different layers, an external serous layer and an internal mucous layer; between the two there develops later a third layer, the vascular (blood—vessel) layer.* (* The technical terms which are bound to creep into this chapter will be fully understood later on.—Translator.)

Karl Ernst von Baer, who had set afoot Pander's investigation, and had shown the liveliest interest in it after Pander's departure from Wurtzburg, began his own much more comprehensive research in 1819. He published the mature result nine years afterwards in his famous work, Animal Embryology: Observation and Reflection (not translated). This classic work still remains a model of careful observation united to profound philosophic speculation. The first part appeared in 1828, the second in 1837. The book proved to be the foundation on which the whole science of embryology has built down to our own day. It so far surpassed its predecessors, and Pander in particular, that it has become, after Wolff's work, the chief base of modern embryology.

Baer was one of the greatest scientists of the nineteenth century, and exercised considerable influence on other branches of biology as well. He built up the theory of germinal layers, as a whole and in detail, so clearly and solidly that it has been the starting—point of embryological research ever since. He taught that in all the vertebrates first two and then four of these germinal layers are formed; and that the earliest rudimentary organs of the body arise by the conversion of these layers into tubes. He described the first appearance of the vertebrate embryo, as it may be seen in the globular yelk of the fertilised egg, as an oval disk which first divides into two layers. From the upper or animal layer are developed all the organs which accomplish the phenomena of animal life—the functions of sensation and motion, and the covering of the body. From the lower or vegetative layer come the organs which effect the vegetative life of the organism—nutrition, digestion, blood—formation, respiration, secretion, reproduction, etc.

Each of these original layers divides, according to Baer, into two thinner and superimposed layers or plates. He calls the two plates of the animal layer, the

41

skin–stratum and muscle–stratum. From the upper of these plates, the skin–stratum, the external skin, or outer covering of the body, the central nervous system, and the sense–organs, are formed. From the lower, or muscle–stratum, the muscles, or fleshy parts and the bony skeleton—in a word, the motor organs—are evolved. In the same way, Baer said, the lower or vegetative layer splits into two plates, which he calls the vascular–stratum and the mucous–stratum. From the outer of the two (the vascular) the heart, blood–vessels, spleen, and the other vascular glands, the kidneys, and sexual glands, are formed. From the fourth or mucous layer, in fine, we get the internal and digestive lining of the alimentary canal and all its dependencies, the liver, lungs, salivary glands, etc. Baer had, in the main, correctly judged the significance of these four secondary embryonic layers, and he followed the conversion of them into the tube–shaped primitive organs with great perspicacity. He first solved the difficult problem of the transformation of this four–fold, flat, leaf–shaped, embryonic disk into the complete vertebrate body, through the conversion of the layers or plates into tubes. The flat leaves bend themselves in obedience to certain laws of growth; the borders of the curling plates approach nearer and nearer; until at last they come into actual contact. Thus out of the flat gut–plate is formed a hollow gut–tube, out of the flat spinal plate a hollow nerve–tube, from the skin–plate a skin–tube, and so on.

Among the many great services which Baer rendered to embryology, especially vertebrate embryology, we must not forget his discovery of the human ovum. Earlier scientists had, as a rule, of course, assumed that man developed out of an egg, like the other animals. In fact, the preformation theory held that the germs of the whole of humanity were stored already in Eve's ova. But the real ovum escaped detection until the year 1827. This ovum is extremely small, being a tiny round vesicle about the 1/120 of an inch in diameter; it can be seen under very favourable circumstances with the naked eye as a tiny particle, but is otherwise quite invisible. This particle is formed in the ovary inside a much larger globule, which takes the name of the Graafian follicle, from its discoverer, Graaf, and had previously been regarded as the true ovum. However, in 1827 Baer proved that it was not the real ovum, which is much smaller, and is contained within the follicle. (Compare the end of Chapter 2.29.)

Baer was also the first to observe what is known as the segmentation sphere of the vertebrate; that is to say, the round vesicle which first develops out of the impregnated ovum, and the thin wall of which is made up of a single layer of regular, polygonal (many–cornered) cells (see the illustration in Chapter 1.12). Another discovery of his that was of great importance in constructing the vertebrate stem and the characteristic organisation of this extensive group (to which man belongs) was the detection of the axial rod, or the chorda dorsalis. There is a long, round, cylindrical rod of cartilage which runs down the longer axis of the vertebrate embryo; it appears at an early stage, and is the first sketch of the spinal column, the solid skeletal axis of the vertebrate. In the lowest of the vertebrates, the amphioxus, the internal skeleton consists only of this cord throughout life. But even in the case of man and all the higher vertebrates it is round this cord that the spinal column and the brain are afterwards formed.

However, important as these and many other discoveries of Baer's were in vertebrate embryology, his researches were even more influential, from the circumstance that he was the first to employ the comparative method in studying the development of the animal frame. Baer occupied himself chiefly with the embryology of vertebrates (especially the birds and fishes). But he by no means confined his attention to these, gradually taking the various groups of the invertebrates into his sphere of study. As the general result of his comparative embryological research, Baer distinguished four different modes of development and four corresponding groups in the animal world. These chief groups or types are: 1, the vertebrata; 2, the articulata; 3, the mollusca; and 4, all the lower groups which were then wrongly comprehended under the general name of the radiata. Georges Cuvier had been the first to formulate this distinction, in 1812. He showed that these groups present specific differences in their whole internal structure, and the connection and disposal of their systems of organs; and that, on the other hand, all the animals of the same type—say, the vertebrates—essentially agreed in their inner structure, in spite of the greatest superficial differences. But Baer proved that these four groups are also quite differently developed from the ovum; and that the series of embryonic forms is the same throughout for animals of the same type, but different in the case of other animals. Up to that time the chief aim in the classification of the animal kingdom was to arrange all the animals from lowest to highest, from the infusorium to man, in one long and continuous series. The erroneous idea prevailed nearly everywhere that there was one uninterrupted chain of evolution from the lowest animal to the highest. Cuvier and Baer proved that this view was false, and that we must distinguish four totally different types of animals, on the ground of anatomic structure and embryonic development.

Baer's epoch-making works aroused an extraordinary and widespread interest in embryological research. Immediately afterwards we find a great number of observers at work in the newly opened field, enlarging it in a very short time with great energy by their various discoveries in detail. Next to Baer's comes the admirable work of Heinrich Rathke, of Konigsberg (died 1860); he made an extensive study of the embryology, not only of the invertebrates (crustaceans, insects, molluscs), but also, and particularly, of the vertebrates (fishes, tortoises, serpents, crocodiles, etc.). We owe the first comprehensive studies of mammal embryology to the careful research of Wilhelm Bischoff, of Munich; his embryology of the rabbit (1840), the dog (1842), the guinea–pig (1852), and the doe (1854), still form classical studies. About the same time a great impetus was given to the embryology of the invertebrates. The way was opened through this obscure province by the studies of the famous Berlin zoologist, Johannes Muller, on the echinoderms. He was followed by Albert Kolliker, of Wurtzburg, writing on the cuttlefish (or the cephalopods), Siebold and Huxley on worms and zoophytes, Fritz Muller (Desterro) on the crustacea, Weismann on insects, and so on. The number of workers in this field has greatly increased of late, and a quantity of new and astonishing discoveries have been made. One notices, in several of these recent works on embryology, that their authors are too little acquainted with comparative anatomy and classification. Palaeontology is, unfortunately, altogether neglected by many of these new workers, although this interesting science furnishes most important facts for phylogeny, and thus often proves of very great service in ontogeny.

A very important advance was made in our science in 1839, when the cellular theory was established, and a new field of inquiry bearing on embryology was suddenly opened. When the famous botanist, M. Schleiden, of Jena, showed in 1838, with the aid of the microscope, that every plant was made up of innumerable elementary parts, which we call cells, a pupil of Johannes Muller at Berlin, Theodor Schwann, applied the discovery at once to the animal organism. He showed that in the animal body as well, when we examine its tissues in the microscope, we find these cells everywhere to be the elementary units. All the different tissues of the organism, especially the very dissimilar tissues of the nerves, muscles, bones, external skin, mucous lining, etc., are originally formed out of cells; and this is also true of all the tissues of the plant. These cells are separate living beings; they are the citizens of the State which the entire multicellular organism seems to be. This important discovery was bound to be of service to embryology, as it raised a number of new questions. What is the relation of the cells to the germinal layers? Are the germinal layers composed of cells, and what is their relation to the cells of the tissues that form later? How does the ovum stand in the cellular theory? Is the ovum itself a cell, or is it composed of cells? These important questions were now imposed on the embryologist by the cellular theory.

The most notable effort to answer these questions—which were attacked on all sides by different students—is contained in the famous work, Inquiries into the Development of the Vertebrates (not translated) of Robert Remak, of Berlin (1851). This gifted scientist succeeded in mastering, by a complete reform of the science, the great difficulties which the cellular theory had at first put in the way of embryology. A Berlin anatomist, Carl Boguslaus Reichert, had already attempted to explain the origin of the tissues. But this attempt was bound to miscarry, since its not very clear–headed author lacked a sound acquaintance with embryology and the cell theory, and even with the structure and development of the tissue in particular. Remak at length brought order into the dreadful confusion that Reichert had caused; he gave a perfectly simple explanation of the origin of the tissues. In his opinion the animal ovum is always a simple cell: the germinal layers which develop out of it are always composed of cells; and these cells that constitute the germinal layers arise simply from the continuous and repeated cleaving (segmentation) of the original solitary cell. It first divides into two and then into four cells; out of these four cells are born eight, then sixteen, thirty–two, and so on. Thus, in the embryonic development of every animal and plant there is formed first of all out of the simple egg cell, by a repeated subdivision, a cluster of cells, as Kolliker had already stated in connection with the cephalopods in 1844. The cells of this group spread themselves out flat and form leaves or plates; each of these leaves is formed exclusively out of cells. The cells of different layers assume different shapes, increase, and differentiate; and in the end there is a further cleavage (differentiation) and division of work of the cells within the layers, and from these all the different tissues of the body proceed.

These are the simple foundations of histogeny, or the science that treats of the development of the tissues (hista), as it was established by Remak and Kolliker. Remak, in determining more closely the part which the different germinal layers play in the formation of the various tissues and organs, and in applying the theory of evolution to the cells and the tissues they compose, raised the theory of germinal layers, at least as far as it regards the vertebrates, to a high degree of perfection.

Remak showed that three layers are formed out of the two germinal layers which compose the first simple leaf–shaped structure of the vertebrate body (or the "germinal disk"), as the lower layer splits into two plates. These three layers have a very definite relation to the various tissues. First of all, the cells which form the outer skin of the body (the epidermis), with its various dependencies (hairs, nails, etc.)—that is to say, the entire outer envelope of the body—are developed out of the outer or upper layer; but there are also developed in a curious way out of the same layer the cells which form the central nervous system, the brain and the spinal cord. In the second place, the inner or lower germinal layer gives rise only to the cells which form the epithelium (the whole inner lining) of the alimentary canal and all that depends on it (the lungs, liver, pancreas, etc.), or the tissues that receive and prepare the nourishment of the body. Finally, the middle layer gives rise to all the other tissues of the body, the muscles, blood, bones, cartilage, etc. Remak further proved that this middle layer, which he calls "the motor–germinative layer," proceeds to subdivide into two secondary layers. Thus we find once more the four layers which Baer had indicated. Remak calls the outer secondary leaf of the middle layer (Baer's "muscular layer") the "skin layer" (it would be better to say, skin–fibre layer); it forms the outer wall of the body (the true skin, the muscles, etc.). To the inner secondary leaf (Baer's "vascular layer") he gave the name of the "alimentary–fibre layer"; this forms the outer envelope of the alimentary canal, with the mesentery, the heart, the blood–vessels, etc.

On this firm foundation provided by Remak for histogeny, or the science of the formation of the tissues, our knowledge has been gradually built up and enlarged in detail. There have been several attempts to restrict and even destroy Remak's principles. The two anatomists, Reichert (of Berlin) and Wilhelm His (of Leipzic), especially, have endeavoured in their works to introduce a new conception of the embryonic development of the vertebrate, according to which the two primary germinal layers would not be the sole sources of formation. But these efforts were so seriously marred by ignorance of comparative anatomy, an imperfect acquaintance with ontogenesis, and a complete neglect of phylogenesis, that they could not have more than a passing success. We can only explain how these curious attacks of Reichert and His came to be regarded for a time as advances by the general lack of discrimination and of grasp of the true object of embryology.

Wilhelm His published, in 1868, his extensive Researches into the Earliest Form of the Vertebrate Body,* (* None of His's works have been translated into English.) one of the curiosities of embryological literature. The author imagines that he can build a "mechanical theory of embryonic development" by merely giving an exact description of the embryology of the chick, without any regard to comparative anatomy and phylogeny, and thus falls into an error that is almost without parallel in the history of biological literature. As the final result of his laborious investigations, His tells us "that a comparatively simple law of growth is the one essential thing in the first development. Every formation, whether it consist in cleavage of layers, or folding, or complete division, is a consequence of this fundamental law." Unfortunately, he does not explain what this "law of growth" is; just as other opponents of the theory of selection, who would put in its place a great "law of evolution," omit to tell us anything about the nature of this. Nevertheless, it is quite clear from His's works that he imagines constructive Nature to be a sort of skilful tailor. The ingenious operator succeeds in bringing into existence, by "evolution," all the various forms of living things by cutting up in different ways the germinal layers, bending and folding, tugging and splitting, and so on.

His's embryological theories excited a good deal of interest at the time of publication, and have evoked a fair amount of literature in the last few decades. He professed to explain the most complicated parts of organic construction (such as the development of the brain) in the simplest way on mechanical principles, and to derive them immediately from simple physical processes (such as unequal distribution of strain in an elastic plate). It is quite true that a mechanical or monistic explanation (or a reduction of natural processes) is the ideal of modern science, and this ideal would be realised if we could succeed in expressing these formative processes in mathematical formulae. His has, therefore, inserted plenty of numbers and measurements in his embryological works, and given them an air of "exact" scholarship by putting in a quantity of mathematical tables. Unfortunately, they are of no value, and do not help us in the least in forming an "exact" acquaintance with the embryonic phenomena. Indeed, they wander from the true path altogether by neglecting the phylogenetic method; this, he thinks, is "a mere by–path," and is "not necessary at all for the explanation of the facts of embryology," which are the direct consequence of physiological principles. What His takes to be a simple physical process—for instance, the folding of the germinal layers (in the formation of the medullary tube, alimentary tube, etc.)—is, as a matter of fact, the direct result of the growth of the various cells which form those organic structures; but these growth–motions have themselves been transmitted by heredity from parents and ancestors, and are only the hereditary repetition of countless phylogenetic changes which have taken place for thousands of years in the race–history of the said ancestors. Each of these historical changes was, of course, originally due to adaptation; it was, in other words, physiological, and reducible to mechanical causes. But we have, naturally, no means of observing them now. It is only by the hypotheses of the science of evolution that we can form an approximate idea of the organic links in this historic chain.

All the best recent research in animal embryology has led to the confirmation and development of Baer and Remak's theory of the germinal layers. One of the most important advances in this direction of late was the discovery that the two primary layers out of which is built the body of all vertebrates (including man) are also present in all the invertebrates, with the sole exception of the lowest group, the unicellular protozoa. Huxley had detected them in the medusa in 1849. He showed that the two layers of cells from which the body of this zoophyte is developed correspond, both morphologically and physiologically, to the two original germinal layers of the vertebrate. The outer layer, from which come the external skin and the muscles, was then called by Allman (1853) the "ectoderm" (outer layer, or skin); the inner layer, which forms the alimentary and reproductory organs, was called the "entoderm" (= inner layer). In 1867 and the following years the discovery of the germinal layers was extended to other groups of the invertebrates. In particular, the indefatigable Russian zoologist, Kowalevsky, found them in all the most diverse sections of the invertebrates—the worms, tunicates, echinoderms, molluscs, articulates, etc.

In my monograph on the sponges (1872) I proved that these two primary germinal layers are also found in that group, and that they may be traced from it right up to man, through all the various classes, in identical form. This "homology of the two primary germinal layers" extends through the whole of the metazoa, or tissue–forming animals; that is to say, through the whole animal kingdom, with the one exception of its lowest section, the unicellular beings, or protozoa. These lowly organised animals do not form germinal layers, and therefore do not succeed in forming true tissue. Their whole body consists of a single cell (as is the case with the amoebae and infusoria), or of a loose aggregation of only slightly differentiated cells, though it may not even reach the full structure of a single cell (as with the monera). But in all other animals the ovum first grows into two primary layers, the outer or animal layer (the ectoderm, epiblast, or ectoblast), and the inner or vegetal layer (the entoderm, hypoblast, or endoblast); and from these the tissues and organs are formed. The first and oldest organ of all these metazoa is the primitive gut (or progaster) and its opening, the primitive mouth (prostoma). The typical embryonic form of the metazoa, as it is presented for a time by this simple structure of the two–layered body, is called the gastrula; it is to be conceived as the hereditary reproduction of some primitive common ancestor of the metazoa, which we call the gastraea. This applies to the sponges and other zoophyta, and to the worms, the mollusca, echinoderma, articulata, and vertebrata. All these animals may be comprised under the general heading of "gut animals," or metazoa, in contradistinction to the gutless protozoa.

I have pointed out in my Study of the Gastraea Theory [not translated] (1873) the important consequences of this conception in the morphology and classification of the animal world. I also divided the realm of metazoa into two great groups, the lower and higher metazoa. In the first are comprised the coelenterata (also called zoophytes, or plant–animals). In the lower forms of this group the body consists throughout life merely of the primary germinal layers, with the cells sometimes more and sometimes less differentiated. But with the higher forms of the coelentarata (the corals, higher medusae, ctenophorae, and platodes) a middle layer, or mesoderm, often of considerable size, is developed between the other two layers; but blood and an internal cavity are still lacking.

To the second great group of the metazoa I gave the name of the coelomaria, or bilaterata (or the bilateral higher forms). They all have a cavity within the body (coeloma), and most of them have blood and blood–vessels. In this are comprised the six higher stems of the animal kingdom, the annulata and their descendants, the mollusca, echinoderma, articulata, tunicata, and vertebrata. In all these bilateral organisms the two–sided body is formed out of four secondary germinal layers, of which the inner two construct the wall of the alimentary canal, and the outer two the wall of the body. Between the two pairs of layers lies the cavity (coeloma).

Although I laid special stress on the great morphological importance of this cavity in my Study of the Gastraea Theory, and endeavoured to prove the significance of the four secondary germinal layers in the organisation of the coelomaria, I was unable to deal satisfactorily with the difficult question of the mode of their origin. This was done eight years afterwards by the brothers Oscar and Richard Hertwig in their careful and extensive comparative studies. In their masterly Coelum Theory: An Attempt to Explain the Middle Germinal Layer [not translated] (1881) they showed that in most of the metazoa, especially in all the vertebrates, the body–cavity arises in the same way, by the outgrowth of two sacs from the inner layer. These two coelom–pouches proceed from the rudimentary mouth of the gastrula, between the two primary layers. The inner plate of the two–layered coelom–pouch (the visceral layer) joins itself to the entoderm; the outer plate (parietal layer) unites with the ectoderm. Thus are formed the double–layered gut–wall within and the double–layered body–wall without; and between the two is formed the cavity of the coelom, by the blending of the right and left coelom–sacs. We shall see this more fully in Chapter 1.10.

The many new points of view and fresh ideas suggested by my gastraea theory and Hertwig's coelom theory led to the publication of a number of writings on the theory of germinal layers. Most of them set out to oppose it at first, but in the end the majority supported it. Of late years both theories are accepted in their essential features by nearly every competent man of science, and light and order have been introduced into this once dark and contradictory field of research. A further cause of congratulation for this solution of the great embryological controversy is that it brought with it a recognition of the need for phylogenetic study and explanation.

Interest and practice in embryological research have been remarkably stimulated during the past thirty years by this appreciation of phylogenetic methods. Hundreds of assiduous and able observers are now engaged in the development of comparative embryology and its establishment on a basis of evolution, whereas they numbered only a few dozen not many decades ago. It would take too long to enumerate even the most important of the countless valuable works which have enriched embryological literature since that time. References to them will be found in the latest manuals of embryology of Kolliker, Balfour, Hertwig, Kollman, Korschelt, and Heider.

Kolliker's Entwickelungsgeschichte des Menschen und der hoherer Thiere, the first edition of which appeared forty-two years ago, had the rare merit at that time of gathering into presentable form the scattered attainments of the science, and expounding them in some sort of unity on the basis of the cellular theory and the theory of germinal layers. Unfortunately, the distinguished Wurtzburg anatomist, to whom comparative anatomy, histology, and ontogeny owe so much, is opposed to the theory of descent generally and to Darwinism in particular. All the other manuals I have mentioned take a decided stand on evolution. Francis Balfour has carefully collected and presented with discrimination, in his Manual of Comparative Embryology (1880), the very scattered and extensive literature of the subject; he has also widened the basis of the gastraea theory by a comparative description of the rise of the organs from the germinal layers in all the chief groups of the animal kingdom, and has given a most thorough empirical support to the principles I have formulated. A comparison of his work with the excellent Text-book of the Embryology of the Vertebrates (1890) [translation 1895] of Korschelt and Heider shows what astonishing progress has been made in the science in the course of ten years. I would especially recommend the manuals of Julius Kollmann and Oscar Hertwig to those readers who are stimulated to further study by these chapters on human embryology. Kollmann's work is commendable for its clear treatment of the subject and very fine original illustrations; its author adheres firmly to the biogenetic law, and uses it throughout with considerable profit. That is not the case in Oscar Hertwig's recent Text–book of the Embryology of Man and the Mammals [translations 1892 and 1899] (seventh edition 1902). This able anatomist has of late often been quoted as an opponent of the biogenetic law, although he himself had demonstrated its great value thirty years ago. His recent vacillation is partly due to the timidity which our "exact" scientists have with regard to hypotheses; though it is impossible to make any headway in the explanation of facts without them. However, the purely descriptive part of embryology in Hertwig's Text–book is very thorough and reliable.

A new branch of embryological research has been studied very assiduously in the last decade of the nineteenth century—namely, "experimental embryology." The great importance which has been attached to the application of physical experiments to the living organism for the last hundred years, and the valuable results that it has given to physiology in the study of the vital phenomena, have led to its extension to embryology. I was the first to make experiments of this kind during a stay of four months on the Canary Island, Lanzerote, in 1866. I there made a thorough investigation of the almost unknown embryology of the siphonophorae. I cut a number of the embryos of these animals (which develop freely in the water, and pass through a very curious transformation), at an early stage, into several pieces, and found that a fresh organism (more or less complete, according to the size of the piece) was developed from each particle. More recently some of my pupils have made similar experiments with the embryos of vertebrates (especially the frog) and some of the invertebrates. Wilhelm Roux, in particular, has made extensive experiments, and based on them a special "mechanical embryology," which has given rise to a good deal of discussion and controversy. Roux has published a special journal for these subjects since 1895, the Archiv fur Entwickelungsmechanik. The contributions to it are very varied in value. Many of them are valuable papers on the physiology and pathology of the embryo. Pathological experiments—the placing of the embryo in abnormal conditions—have yielded many interesting results; just as the physiology of the normal body has for a long time derived assistance from the pathology of the diseased organism. Other of these mechanical–embryological articles return to the erroneous methods of His, and are only misleading. This must be said of the many contributions of mechanical embryology which take up a position of hostility to the theory of descent and its chief embryological foundation—the biogenetic law. This law, however, when rightly understood, is not opposed to, but is the best and most solid support of, a sound mechanical embryology. Impartial reflection and a due attention to paleontology and comparative anatomy should convince these one-sided mechanicists that the facts they have discovered—and, indeed, the whole embryological process—cannot be fully understood without the theory of descent and the biogenetic law.

CHAPTER 1.4. THE OLDER PHYLOGENY.

The embryology of man and the animals, the history of which we have reviewed in the last two chapters, was mainly a descriptive science forty years ago. The earlier investigations in this province were chiefly directed to the discovery, by careful observation, of the wonderful facts of the embryonic development of the animal body from the ovum. Forty years ago no one dared attack the question of the CAUSES of these phenomena. For fully a century, from the year 1759, when Wolff's solid Theoria generationis appeared, until 1859, when Darwin published his famous Origin of Species, the real causes of the embryonic processes were quite unknown. No one thought of seeking the agencies that effected this marvellous succession of structures. The task was thought to be so difficult as almost to pass beyond the limits of human thought. It was reserved for Charles Darwin to initiate us into the knowledge of these causes. This compels us to recognise in this great genius, who wrought a complete revolution in the whole field of biology, a founder at the same time of a new period in embryology. It is true that Darwin occupied himself very little with direct embryological research, and even in his chief work he only touches incidentally on the embryonic phenomena; but by his reform of the theory of descent and the founding of the theory of selection he has given us the means of attaining to a real knowledge of the causes of embryonic formation. That is, in my opinion, the chief feature in Darwin's incalculable influence on the whole science of evolution.

When we turn our attention to this latest period of embryological research, we pass into the second division of organic evolution—stem-evolution, or phylogeny. I have already indicated in Chapter 1.1 the important and intimate causal connection between these two sections of the science of evolution—between the evolution of the individual and that of his ancestors. We have formulated this connection in the biogenetic law; the shorter evolution, that of the individual, or ontogenesis, is a rapid and summary repetition, a condensed recapitulation, of the larger evolution, or that of the species. In this principle we express all the essential points relating to the causes of evolution; and we shall seek throughout this work to confirm this principle and lend it the support of facts. When we look to its CAUSAL significance, perhaps it would be better to formulate the biogenetic law thus: "The evolution of the species and the stem (phylon) shows us, in the physiological functions of heredity and adaptation, the conditioning causes on which the evolution of the individual depends"; or, more briefly: "Phylogenesis is the mechanical cause of ontogenesis."

But before we examine the great achievement by which Darwin revealed the causes of evolution to us, we must glance at the efforts of earlier scientists to attain this object. Our historical inquiry into these will be even shorter than that into the work done in the field of ontogeny. We have very few names to consider here. At the head of them we find the great French naturalist, Jean Lamarck, who first established evolution as a scientific theory in 1809. Even before his time, however, the chief philosopher, Kant, and the chief poet, Goethe, of Germany had occupied themselves with the subject. But their efforts passed almost without recognition in the eighteenth century. A "philosophy of nature" did not arise until the beginning of the nineteenth century. In the whole of the time before this no one had ventured to raise seriously the question of the origin of species, which is the culminating point of phylogeny. On all sides it was regarded as an insoluble enigma.

The whole science of the evolution of man and the other animals is intimately connected with the question of the nature of species, or with the problem of the origin of the various animals which we group together under the name of species. Thus the definition of the species becomes important. It is well known that this definition was given by Linne, who, in his famous Systema Naturae (1735), was the first to classify and name the various groups of animals and plants, and drew up an orderly scheme of the species then known. Since that time "species" has been the most important and indispensable idea in descriptive natural history, in zoological and botanical classification; although there have been endless controversies as to its real meaning.

What, then, is this "organic species"? Linne himself appealed directly to the Mosaic narrative; he believed that, as it is stated in Genesis, one pair of each species of animals and plants was created in the beginning, and that all the individuals of each species are the descendants of these created couples. As for the hermaphrodites (organisms that have male and female organs in one being), he thought it sufficed to assume the creation of one sole individual, since this would be fully competent to propagate its species. Further developing these mystic ideas, Linne went on to borrow from Genesis the account of the deluge and of Noah's ark as a ground for a science of the geographical and topographical distribution of organisms. He accepted the story that all the plants, animals, and men on the earth were swept away in a universal deluge, except the couples preserved with Noah in the ark, and ultimately landed on Mount Ararat. This mountain seemed to Linne particularly suitable for the landing, as it reaches a height of more than 16,000 feet, and thus provides in its higher zones the several climates demanded by the various species of animals and plants: the animals that were accustomed to a cold climate could remain at the summit; those used to a warm climate could descend to the foot; and those requiring a temperate climate could remain half–way down. From this point the re–population of the earth with animals and plants could proceed.

It was impossible to have any scientific notion of the method of evolution in Linne's time, as one of the chief sources of information, paleontology, was still wholly unknown. This science of the fossil remains of extinct animals and plants is very closely bound up with the whole question of evolution. It is impossible to explain the origin of living organisms without appealing to it. But this science did not rise until a much later date. The real founder of scientific paleontology was Georges Cuvier, the most distinguished zoologist who, after Linne, worked at the classification of the animal world, and effected a complete revolution in systematic zoology at the beginning of the nineteenth century. In regard to the nature of the species he associated himself with Linne and the Mosaic story of creation, though this was more difficult for him with his acquaintance with fossil remains. He clearly showed that a number of quite different animal populations have lived on the earth; and he claimed that we must distinguish a number of stages in the history of our planet, each of which was characterised by a special population of animals and plants. These successive populations were, he said, quite independent of each other, and therefore the supernatural creative act, which was demanded as the origin of the animals and plants by the dominant creed, must have been repeated several times. In this way a whole series of different creative periods must have succeeded each other; and in connection with these he had to assume that stupendous revolutions or cataclysms—something like the legendary deluge—must have taken place repeatedly. Cuvier was all the more interested in these catastrophes or cataclysms as geology was just beginning to assert itself, and great progress was being made in our knowledge of the structure and formation of the earth's crust. The various strata of the crust were being carefully examined, especially by the famous geologist Werner and his school, and the fossils found in them were being classified; and these researches also seemed to point to a variety of creative periods. In each period the earth's crust, composed of the various strata, seemed to be differently constituted, just like the population of animals and plants that then lived on it. Cuvier combined this notion with the results of his own paleontological and zoological research; and in his effort to get a consistent view of the whole process of the earth's history he came to form the theory which is known as "the catastrophic theory," or the theory of terrestrial revolutions. According to this theory, there have been a series of mighty cataclysms on the earth, and these have suddenly destroyed the whole animal and plant population then living on it; after each cataclysm there was a fresh creation of living things throughout the earth. As this creation could not be explained by natural laws, it was necessary to appeal to an intervention on the part of the Creator. This catastrophic theory, which Cuvier described in a special work, was soon generally accepted, and retained its position in biology for half a century.

However, Cuvier's theory was completely overthrown sixty years ago by the geologists, led by Charles Lyell, the most distinguished worker in this field of science. Lyell proved in his famous Principles of Geology (1830) that the theory was false, in so far as it concerned the crust of the earth; that it was totally unnecessary to bring in supernatural agencies or general catastrophes in order to explain the structure and formation of the mountains; and that we can explain them by the familiar agencies which are at work to–day in altering and reconstructing the surface of the earth. These causes are—the action of the atmosphere and water in its various forms (snow, ice, fog, rain, the wear of the river, and the stormy ocean), and the volcanic action which is exerted by the molten central mass. Lyell convincingly proved that these natural causes are quite adequate to explain every feature in the build and formation of the crust. Hence Cuvier's theory of cataclysms was very soon driven out of the province of geology, though it remained for another thirty years in undisputed authority in biology. All the zoologists and botanists who gave any thought to the question of the origin of organisms adhered to Cuvier's erroneous idea of revolutions and new creations.

In order to illustrate the complete stagnancy of biology from 1830 to 1859 on the question of the origin of the various species of animals and plants, I may say, from my own experience, that during the whole of my university studies I never heard a single word said about this most important problem of the science. I was fortunate enough at that time (1852 to 1857) to have the most distinguished masters for every branch of biological science. Not one of them ever mentioned this question of the origin of species. Not a word was ever said about the earlier efforts to understand the formation of living things, nor about Lamarck's Philosophie Zoologique which had made a fresh attack on the problem in 1809. Hence it is easy to understand the enormous opposition that Darwin encountered when he took up the question for the first time. His views seemed to float in the air, without a single previous effort to support them. The whole question of the formation of living things was considered by biologists, until 1859, as pertaining to the province of religion and transcendentalism; even in speculative philosophy, in which the question had been approached from various sides, no one had ventured to give it serious treatment. This was due to the dualistic system of Immanuel Kant, who taught a natural system of evolution as far as the inorganic world was concerned; but, on the whole, adopted a supernaturalist system as regards the origin of living things. He even went so far as to say: "It is quite certain that we cannot even satisfactorily understand, much less explain, the nature of an organism and its internal forces on purely mechanical principles; it is so certain, indeed, that we may confidently say: 'It is absurd for a man to imagine even that some day a Newton will arise who will explain the origin of a single blade of grass by natural laws not controlled by design'—such a hope is entirely forbidden us." In these words Kant definitely adopts the dualistic and teleological point of view for biological science.

Nevertheless, Kant deserted this point of view at times, particularly in several remarkable passages which I have dealt with at length in my Natural History of Creation (chapter 5), where he expresses himself in the opposite, or monistic, sense. In fact, these passages would justify one, as I showed, in claiming his support for the theory of evolution. However, these monistic passages are only stray gleams of light; as a rule, Kant adheres in biology to the obscure dualistic ideas, according to which the forces at work in inorganic nature are quite different from those of the organic world. This dualistic system prevails in academic philosophy to–day—most of our philosophers still regarding these two provinces as totally distinct. They put, on the one side, the inorganic or "lifeless" world, in which there are at work only mechanical laws, acting necessarily and without design; and, on the other, the province of organic nature, in which none of the phenomena can be properly understood, either as regards their inner nature or their origin, except in the light of preconceived design, carried out by final or purposive causes.

The prevalence of this unfortunate dualistic prejudice prevented the problem of the origin of species, and the connected question of the origin of man, from being regarded by the bulk of people as a scientific question at all until 1859. Nevertheless, a few distinguished students, free from the current prejudice, began, at the commencement of the nineteenth century, to make a serious attack on the problem. The merit of this attaches particularly to what is known as "the older school of natural philosophy," which has been so much misrepresented, and which included Jean Lamarck, Buffon, Geoffroy St. Hilaire, and Blainville in France; Wolfgang Goethe, Reinhold Treviranus, Schelling, and Lorentz Oken in Germany [and Erasmus Darwin in England].

The gifted natural philosopher who treated this difficult question with the greatest sagacity and comprehensiveness was Jean Lamarck. He was born at Bazentin, in Picardy, on August 1st, 1744; he was the son of a clergyman, and was destined for the Church. But he turned to seek glory in the army, and eventually devoted himself to science.

His Philosophie Zoologique was the first scientific attempt to sketch the real course of the origin of species, the first "natural history of creation" of plants, animals, and men. But, as in the case of Wolff's book, this remarkably able work had no influence whatever; neither one nor the other could obtain any recognition from their prejudiced contemporaries. No man of science was stimulated to take an interest in the work, and to develop the germs it contained of the most important biological truths. The most distinguished botanists and zoologists entirely rejected it, and did not even deign to reply to it. Cuvier, who lived and worked in the same city, has not thought fit to devote a single syllable to this great achievement in his memoir on progress in the sciences, in which the pettiest observations found a place. In short, Lamarck's Philosophie Zoologique shared the fate of Wolff's theory of development, and was for half a century ignored and neglected. The German scientists, especially Oken and Goethe, who were occupied with similar speculations at the same time, seem to have known nothing about Lamarck's work. If they had known it, they would have been greatly helped by it, and might have carried the theory of evolution much farther than they found it possible to do.

To give an idea of the great importance of the Philosophie Zoologique, I will briefly explain Lamarck's leading thought. He held that there was no essential difference between living and lifeless beings. Nature is one united and connected system of phenomena; and the forces which fashion the lifeless bodies are the only ones at work in the kingdom of living things. We have, therefore, to use the same method of investigation and explanation in both provinces. Life is only a physical phenomenon. All the plants and animals, with man at their head, are to be explained, in structure and life, by mechanical or efficient causes, without any appeal to final causes, just as in the case of minerals and other inorganic bodies. This applies equally to the origin of the various species. We must not assume any original creation, or repeated creations (as in Cuvier's theory), to explain this, but a natural, continuous, and necessary evolution. The whole evolutionary process has been uninterrupted. All the different kinds of animals and plants which we see to–day, or that have ever lived, have descended in a natural way from earlier and different species; all come from one common stock, or from a few common ancestors. These remote ancestors must have been quite simple organisms of the lowest type, arising by spontaneous generation from inorganic matter. The succeeding species have been constantly modified by adaptation to their varying environment (especially by use and habit), and have transmitted their modifications to their successors by heredity.

Lamarck was the first to formulate as a scientific theory the natural origin of living things, including man, and to push the theory to its extreme conclusions—the rise of the earliest organisms by spontaneous generation (or abiogenesis) and the descent of man from the nearest related mammal, the ape. He sought to explain this last point, which is of especial interest to us here, by the same agencies which he found at work in the natural origin of the plant and animal species. He considered use and habit (adaptation) on the one hand, and heredity on the other, to be the chief of these agencies. The most important modifications of the organs of plants and animals are due, in his opinion, to the function of these very organs, or to the use or disuse of them. To give a few examples, the woodpecker and the humming–bird have got their peculiarly long tongues from the habit of extracting their food with their tongues from deep and narrow folds or canals; the frog has developed the web between his toes by his own swimming; the giraffe has lengthened his neck by stretching up to the higher branches of trees, and so on. It is quite certain that this use or disuse of organs is a most important factor in organic development, but it is not sufficient to explain the origin of species.

To adaptation we must add heredity as the second and not less important agency, as Lamarck perfectly recognised. He said that the modification of the organs in any one individual by use or disuse was slight, but that it was increased by accumulation in passing by heredity from generation to generation. But he missed altogether the principle which Darwin afterwards found to be the chief factor in the theory of transformation—namely, the principle of natural selection in the struggle for existence. It was partly owing to his failure to detect this supremely important element, and partly to the poor condition of all biological science at the time, that Lamarck did not succeed in establishing more firmly his theory of the common descent of man and the other animals.

Independently of Lamarck, the older German school of natural philosophy, especially Reinhold Treviranus, in his Biologie (1802), and Lorentz Oken, in his Naturphilosophie

(1809), turned its attention to the problem of evolution about the end of the eighteenth and beginning of the nineteenth century. I have described its work in my History of Creation (chapter 4). Here I can only deal with the brilliant genius whose evolutionary ideas are of special interest—the greatest of German poets, Wolfgang Goethe. With his keen eye for the beauties of nature, and his profound insight into its life, Goethe was early attracted to the study of various natural sciences. It was the favourite occupation of his leisure hours throughout life. He gave particular and protracted attention to the theory of colours. But the most valuable of his scientific studies are those which relate to that "living, glorious, precious thing," the organism. He made profound research into the science of structures or morphology (morphae = forms). Here, with the aid of comparative anatomy, he obtained the most brilliant results, and went far in advance of his time. I may mention, in particular, his vertebral theory of the skull, his discovery of the pineal gland in man, his system of the metamorphosis of plants, etc. These morphological studies led Goethe on to research into the formation and modification of organic structures which we must count as the first germ of the science of evolution. He approaches so near to the theory of descent that we must regard him, after Lamarck, as one of its earliest founders. It is true that he never formulated a complete scientific theory of evolution, but we find a number of remarkable suggestions of it in his splendid miscellaneous essays on morphology. Some of them are really among the very basic ideas of the science of evolution. He says, for instance (1807): "When we compare plants and animals in their most rudimentary forms, it is almost impossible to distinguish between them. But we may say that the plants and animals, beginning with an almost inseparable closeness, gradually advance along two divergent lines, until the plant at last grows in the solid, enduring tree and the animal attains in man to the highest degree of mobility and freedom." That Goethe was not merely speaking in a poetical, but in a literal genealogical, sense of this close affinity of organic forms is clear from other remarkable passages in which he treats of their variety in outward form and unity in internal structure. He believes that every living thing has arisen by the interaction of two opposing formative forces or impulses. The internal or "centripetal" force, the type or "impulse to specification," seeks to maintain the constancy of the specific forms in the succession of generations: this is heredity. The external or "centrifugal" force, the element of variation or "impulse to metamorphosis," is continually modifying the species by changing their environment: this is adaptation. In these significant conceptions Goethe approaches very close to a recognition of the two great mechanical factors which we now assign as the chief causes of the formation of species.

However, in order to appreciate Goethe's views on morphology, one must associate his decidedly monistic conception of nature with his pantheistic philosophy. The warm and keen interest with which he followed, in his last years, the controversies of contemporary French scientists, and especially the struggle between Cuvier and Geoffroy St. Hilaire (see chapter 4 of The History of Creation), is very characteristic. It is also necessary to be familiar with his style and general tenour of thought in order to appreciate rightly the many allusions to evolution found in his writings. Otherwise, one is apt to make serious errors.

He approached so close, at the end of the eighteenth century, to the principles of the science of evolution that he may well be described as the first forerunner of Darwin, although he did not go so far as to formulate evolution as a scientific system, as Lamarck did.

CHAPTER 1.5. THE MODERN SCIENCE OF EVOLUTION.

We owe so much of the progress of scientific knowledge to Darwin's Origin of Species that its influence is almost without parallel in the history of science. The literature of Darwinism grows from day to day, not only on the side of academic zoology and botany, the sciences which were chiefly affected by Darwin's theory, but in a far wider circle, so that we find Darwinism discussed in popular literature with a vigour and zest that are given to no other scientific conception. This remarkable success is due chiefly to two circumstances. In the first place, all the sciences, and especially biology, have made astounding progress in the last half–century, and have furnished a very vast quantity of proofs of the theory of evolution. In striking contrast to the failure of Lamarck and the older scientists to attract attention to their effort to explain the origin of living things and of man, we have this second and successful effort of Darwin, which was able to gather to its support a large number of established facts. Availing himself of the progress already made, he had very different scientific proofs to allege than Lamarck, or St. Hilaire, or Goethe, or Treviranus had had. But, in the second place, we must acknowledge that Darwin had the special distinction of approaching the subject from an entirely new side, and of basing the theory of descent on a consistent system, which now goes by the name of Darwinism.

Lamarck had unsuccessfully attempted to explain the modification of organisms that descend from a common form chiefly by the action of habit and the use of organs, though with the aid of heredity. But Darwin's success was complete when he independently sought to give a mechanical explanation, on a quite new ground, of this modification of plant and animal structures by adaptation and heredity. He was impelled to his theory of selection on the following grounds. He compared the origin of the various kinds of animals and plants which we modify artificially—by the action of artificial selection in horticulture and among domestic animals—with the origin of the species of animals and plants in their natural state. He then found that the agencies which we employ in the modification of forms by artificial selection are also at work in Nature. The chief of these agencies he held to be "the struggle for life." The gist of this peculiarly Darwinian idea is given in this formula: The struggle for existence produces new species without premeditated design in the life of Nature, in the same way that the will of man consciously selects new races in artificial conditions. The gardener or the farmer selects new forms as he wills for his own profit, by ingeniously using the agency of heredity and adaptation for the modification of structures; so, in the natural state, the struggle for life is always unconsciously modifying the various species of living things. This struggle for life, or competition of organisms in securing the means of subsistence, acts without any conscious design, but it is none the less effective in modifying structures. As heredity and adaptation enter into the closest reciprocal action under its influence, new structures, or alterations of structure, are produced; and these are purposive in the sense that they serve the organism when formed, but they were produced without any pre–conceived aim.

This simple idea is the central thought of Darwinism, or the theory of selection. Darwin conceived this idea at an early date, and then, for more than twenty years, worked at the collection of empirical evidence in support of it before he published his theory. His grandfather, Erasmus Darwin, was an able scientist of the older school of natural philosophy, who published a number of natural–philosophic works about the end of the eighteenth century. The most important of them is his Zoonomia, published in 1794, in which he expounds views similar to those of Goethe and Lamarck, without really knowing anything of the work of these contemporaries. However, in the writings of the grandfather the plastic imagination rather outran the judgment, while in Charles Darwin the two were better balanced.

Darwin did not publish any account of his theory until 1858, when Alfred Russel Wallace, who had independently reached the same theory of selection, published his own work. In the following year appeared the Origin of Species, in which he develops it at length and supports it with a mass of proof. Wallace had reached the same conclusion, but he had not so clear a perception as Darwin of the effectiveness of natural selection in forming species, and did not develop the theory so fully. Nevertheless, Wallace's writings, especially those on mimicry, etc., and an admirable work on The Geographical Distribution of Animals, contain many fine original contributions to the theory of selection. Unfortunately, this gifted scientist has since devoted himself to spiritism.* (* Darwin and Wallace arrived at the theory quite independently. Vide Wallace's Contributions to the Theory of Natural Selection (1870) and Darwinism (1891).)

Darwin's Origin of Species had an extraordinary influence, though not at first on the experts of the science. It took zoologists and botanists several years to recover from the astonishment into which they had been thrown through the revolutionary idea of the work. But its influence on the special sciences with which we zoologists and botanists are concerned has increased from year to year; it has introduced a most healthy fermentation in every branch of biology, especially in comparative anatomy and ontogeny, and in zoological and botanical classification. In this way it has brought about almost a revolution in the prevailing views.

However, the point which chiefly concerns us here—the extension of the theory to man—was not touched at all in Darwin's first work in 1859. It was believed for several years that he had no thought of applying his principles to man, but that he shared the current idea of man holding a special position in the universe. Not only ignorant laymen (especially several theologians), but also a number of men of science, said very naively that Darwinism in itself was not to be opposed; that it was quite right to use it to explain the origin of the various species of plants and animals, but that it was totally inapplicable to man.

In the meantime, however, it seemed to a good many thoughtful people, laymen as well as scientists, that this was wrong; that the descent of man from some other animal species, and immediately from some ape–like mammal, followed logically and necessarily from Darwin's reformed theory of evolution. Many of the acuter opponents of the theory saw at once the justice of this position, and, as this consequence was intolerable, they wanted to get rid of the whole theory.

The first scientific application of the Darwinian theory to man was made by Huxley, the greatest zoologist in England. This able and learned scientist, to whom zoology owes much of its progress, published in 1863 a small work entitled Evidence as to Man's Place in Nature. In the extremely important and interesting lectures which made up this work he proved clearly that the descent of man from the ape followed necessarily from the theory of descent. If that theory is true, we are bound to conceive the animals which most closely resemble man as those from which humanity has been gradually evolved. About the same time Carl Vogt published a larger work on the same subject. We must also mention Gustav Jaeger and Friedrich Rolle among the zoologists who accepted and taught the theory of evolution immediately after the publication of Darwin's book, and maintained that the descent of man from the lower animals logically followed from it. The latter published, in 1866, a work on the origin and position of man.

About the same time I attempted, in the second volume of my General Morphology (1866), to apply the theory of evolution to the whole organic kingdom, including man.* (* Huxley spoke of this "as one of the greatest scientific works ever published."—Translator.) I endeavoured to sketch the probable ancestral trees of the various classes of the animal world, the protists, and the plants, as it seemed necessary to do on Darwinian principles, and as we can actually do now with a high degree of confidence. If the theory of descent, which Lamarck first clearly formulated and Darwin thoroughly established, is true, we should be able to draw up a natural classification of plants and animals in the light of their genealogy, and to conceive the large and small divisions of the system as the branches and twigs of an ancestral tree. The eight genealogical tables which I inserted in the second volume of the General Morphology are the first sketches of their kind. In Chapter 2.27, particularly, I trace the chief stages in man's ancestry, as far as it is possible to follow it through the vertebrate stem. I tried especially to determine, as well as one could at that time, the position of man in the classification of the mammals and its genealogical significance. I have greatly improved this attempt, and treated it in a more popular form, in chapters 26 to 28 of my History of Creation (1868).* (* Of which Darwin said that the Descent of Man would probably never have been written if he had seen it earlier.—Translator.)

It was not until 1871, twelve years after the appearance of The Origin of Species, that Darwin published the famous work which made the much–contested application of his theory to man, and crowned the splendid structure of his system. This important work was The Descent of Man, and Selection in Relation to Sex. In this Darwin expressly drew the conclusion, with rigorous logic, that man also must have been developed out of lower species, and described the important part played by sexual selection in the elevation of man and the other higher animals. He showed that the careful selection which the sexes exercise on each other in regard to sexual relations and procreation, and the aesthetic feeling which the higher animals develop through this, are of the utmost importance in the progressive development of forms and the differentiation of the sexes. The males choosing the handsomest females in one class of animals, and the females choosing only the finest–looking males in another, the special features and the sexual characteristics are increasingly accentuated. In fact, some of the higher animals develop in this connection a finer taste and judgment than man himself. But, even as regards man, it is to this sexual selection that we owe the family–life, which is the chief foundation of civilisation. The rise of the human race is due for the most part to the advanced sexual selection which our ancestors exercised in choosing their mates.

Darwin accepted in the main the general outlines of man's ancestral tree, as I gave it in the General Morphology and the History of Creation, and admitted that his studies led him to the same conclusion. That he did not at once apply the theory to man in his first work was a commendable piece of discretion; such a sequel was bound to excite the strongest opposition to the whole theory. The first thing to do was to establish it as regards the animal and plant worlds. The subsequent extension to man was bound to be made sooner or later.

It is important to understand this very clearly. If all living things come from a common root, man must be included in the general scheme of evolution. On the other hand, if the various species were separately created, man, too, must have been created, and not evolved. We have to choose between these two alternatives. This cannot be too frequently or too strongly emphasised. EITHER all the species of animals and plants are of supernatural origin—created, not evolved—and in that case man also is the outcome of a creative act, as religion teaches, OR the different species have been evolved from a few common, simple ancestral forms, and in that case man is the highest fruit of the tree of evolution.

We may state this briefly in the following principle—The descent of man from the lower animals is a special deduction which inevitably follows from the general inductive law of the whole theory of evolution. In this principle we have a clear and plain statement of the matter. Evolution is in reality nothing but a great induction, which we are compelled to make by the comparative study of the most important facts of morphology and physiology. But we must draw our conclusion according to the laws of induction, and not attempt to determine scientific truths by direct measurement and mathematical calculation. In the study of living things we can scarcely ever directly and fully, and with mathematical accuracy, determine the nature of phenomena, as is done in the simpler study of the inorganic world—in chemistry, physics, mineralogy, and astronomy. In the latter, especially, we can always use the simplest and absolutely safest method—that of mathematical determination. But in biology this is quite impossible for various reasons; one very obvious reason being that most of the facts of the science are very complicated and much too intricate to allow a direct mathematical analysis. The greater part of the phenomena that biology deals with are complicated HISTORICAL PROCESSES, which are related to a far–reaching past, and as a rule can only be approximately estimated. Hence we have to proceed by INDUCTION—that is to say, to draw general conclusions, stage by stage, and with proportionate confidence, from the accumulation of detailed observations. These inductive conclusions cannot command absolute confidence, like mathematical axioms; but they approach the truth, and gain increasing probability, in proportion as we extend the basis of observed facts on which we build. The importance of these inductive laws is not diminished from the circumstance that they are looked upon merely as temporary acquisitions of science, and may be improved to any extent in the progress of scientific knowledge. The same may be said of the attainments of many other sciences, such as geology or archeology. However much they may be altered and improved in detail in the course of time, these inductive truths may retain their substance unchanged.

Now, when we say that the theory of evolution in the sense of Lamarck and Darwin is an inductive law—in fact, the greatest of all biological inductions—we rely, in the first place, on the facts of paleontology. This science gives us some direct acquaintance with the historical phenomena of the changes of species. From the situations in which we find the fossils in the various strata of the earth we gather confidently, in the first place, that the living population of the earth has been gradually developed, as clearly as the earth's crust itself; and that, in the second place, several different populations have succeeded each other in the various geological periods. Modern geology teaches that the formation of the earth has been gradual, and unbroken by any violent revolutions. And when we compare together the various kinds of animals and plants which succeed each other in the history of our planet, we find, in the first place, a constant and gradual increase in the number of species from the earliest times until the present day; and, in the second place, we notice that the forms in each great group of animals and plants also constantly improve as the ages advance. Thus, of the vertebrates there are at first only the lower fishes; then come the higher fishes, and later the amphibia. Still later appear the three higher classes of vertebrates—the reptiles, birds, and mammals, for the first time; only the lowest and least perfect forms of the mammals are found at first; and it is only at a very late period that placental mammals appear, and man belongs to the latest and youngest branch of these. Thus perfection of form increases as well as variety from the earliest to the latest stage. That is a fact of the greatest importance. It can only be explained by the theory of evolution, with which it is in perfect harmony. If the different groups of plants and animals do really descend from each other, we must expect to find this increase in their number and perfection under the influence of natural selection, just as the succession of fossils actually discloses it to us.

Comparative anatomy furnishes a second series of facts which are of great importance for the forming of our inductive law. This branch of morphology compares the adult structures of living things, and seeks in the great variety of organic forms the stable and simple law of organisation, or the common type or structure. Since Cuvier founded this science at the beginning of the nineteenth century it has been a favourite study of the most distinguished scientists. Even before Cuvier's time Goethe had been greatly stimulated by it, and induced to take up the study of morphology. Comparative osteology, or the philosophic study and comparison of the bony skeleton of the vertebrates—one of its most interesting sections—especially fascinated him, and led him to form the theory of the skull which I mentioned before. Comparative anatomy shows that the internal structure of the animals of each stem and the plants of each class is the same in its essential features, however much they differ in external appearance. Thus man has so great a resemblance in the chief features of his internal organisation to the other mammals that no comparative anatomist has ever doubted that he belongs to this class. The whole internal structure of the human body, the arrangement of its various systems of organs, the distribution of the bones, muscles, blood–vessels, etc., and the whole structure of these organs in the larger and the finer scale, agree so closely with those of the other mammals (such as the apes, rodents, ungulates, cetacea, marsupials, etc.) that their external differences are of no account whatever. We learn further from comparative anatomy that the chief features of animal structure are so similar in the various classes (fifty to sixty in number altogether) that they may all be comprised in from eight to twelve great groups. But even in these groups, the stem–forms or animal types, certain organs (especially the alimentary canal) can be proved to have been originally the same for all. We can only explain by the theory of evolution this essential unity in internal structure of all these animal forms that differ so much in outward appearance. This wonderful fact can only be really understood and explained when we regard the internal resemblance as an inheritance from common–stem forms, and the external differences as the effect of adaptation to different environments.

In recognising this, comparative anatomy has itself advanced to a higher stage. Gegenbaur, the most distinguished of recent students of this science, says that with the theory of evolution a new period began in comparative anatomy, and that the theory in turn found a touch stone in the science. "Up to now there is no fact in comparative anatomy that is inconsistent with the theory of evolution; indeed, they all lead to it. In this way the theory receives back from the science all the service it rendered to its method." Until then students had marvelled at the wonderful resemblance of living things in their inner structure without being able to explain it. We are now in a position to explain the causes of this, by showing that this remarkable agreement is the necessary consequence of the inheriting of common stem–forms; while the striking difference in outward appearance is a result of adaptation to changes of environment. Heredity and adaptation alone furnish the true explanation.

But one special part of comparative anatomy is of supreme interest and of the utmost philosophic importance in this connection. This is the science of rudimentary or useless organs; I have given it the name of "dysteleology" in view of its philosophic consequences. Nearly every organism (apart from the very lowest), and especially every highly–developed animal or plant, including man, has one or more organs which are of no use to the body itself, and have no share in its functions or vital aims. Thus we all have, in various parts of our frame, muscles which we never use, as, for instance, in the shell of the ear and adjoining parts. In most of the mammals, especially those with pointed ears, these internal and external ear–muscles are of great service in altering the shell of the ear, so as to catch the waves of sound as much as possible. But in the case of man and other short–eared mammals these muscles are useless, though they are still present. Our ancestors having long abandoned the use of them, we cannot work them at all to–day. In the inner corner of the eye we have a small crescent–shaped fold of skin; this is the last relic of a third inner eye–lid, called the nictitating (winking) membrane. This membrane is highly developed and of great service in some of our distant relations, such as fishes of the shark type and several other vertebrates; in us it is shrunken and useless. In the intestines we have a process that is not only quite useless, but may be very harmful—the vermiform appendage. This small intestinal appendage is often the cause of a fatal illness. If a cherry–stone or other hard body is unfortunately squeezed through its narrow aperture during digestion, a violent inflammation is set up, and often proves fatal. This appendix has no use whatever now in our frame; it is a dangerous relic of an organ that was much larger and was of great service in our vegetarian ancestors. It is still large and important in many vegetarian animals, such as apes and rodents.

There are similar rudimentary organs in all parts of our body, and in all the higher animals. They are among the most interesting phenomena to which comparative anatomy introduces us; partly because they furnish one of the clearest proofs of evolution, and partly because they most strikingly refute the teleology of certain philosophers. The theory of evolution enables us to give a very simple explanation of these phenomena.

We have to look on them as organs which have fallen into disuse in the course of many generations. With the decrease in the use of its function, the organ itself shrivels up gradually, and finally disappears. There is no other way of explaining rudimentary organs. Hence they are also of great interest in philosophy; they show clearly that the monistic or mechanical view of the organism is the only correct one, and that the dualistic or teleological conception is wrong. The ancient legend of the direct creation of man according to a pre-conceived plan and the empty phrases about "design" in the organism are completely shattered by them. It would be difficult to conceive a more thorough refutation of teleology than is furnished by the fact that all the higher animals have these rudimentary organs.

The theory of evolution finds its broadest inductive foundation in the natural classification of living things, which arranges all the various forms in larger and smaller groups, according to their degree of affinity. These groupings or categories of classification—the varieties, species, genera, families, orders, classes, etc.—show such constant features of coordination and subordination that we are bound to look on them as genealogical, and represent the whole system in the form of a branching tree. This is the genealogical tree of the variously related groups; their likeness in form is the expression of a real affinity. As it is impossible to explain in any other way the natural tree-like form of the system of organisms, we must regard it at once as a weighty proof of the truth of evolution. The careful construction of these genealogical trees is, therefore, not an amusement, but the chief task of modern classification.

Among the chief phenomena that bear witness to the inductive law of evolution we have the geographical distribution of the various species of animals and plants over the surface of the earth, and their topographical distribution on the summits of mountains and in the depths of the ocean. The scientific study of these features—the "science of distribution," or chorology (chora = a place)—has been pursued with lively interest since the discoveries made by Alexander von Humboldt. Until Darwin's time the work was confined to the determination of the facts of the science, and chiefly aimed at settling the spheres of distribution of the existing large and small groups of living things. It was impossible at that time to explain the causes of this remarkable distribution, or the reasons why one group is found only in one locality and another in a different place, and why there is this manifold distribution at all. Here, again, the theory of evolution has given us the solution of the problem. It furnishes the only possible explanation when it teaches that the various species and groups of species descend from common stem–forms, whose ever–branching offspring have gradually spread themselves by migration over the earth. For each group of species we must admit a "centre of production," or common home; this is the original habitat in which the ancestral form was developed, and from which its descendants spread out in every direction. Several of these descendants became in their turn the stem–forms for new groups of species, and these also scattered themselves by active and passive migration, and so on. As each migrating organism found a different environment in its new home, and adapted itself to it, it was modified, and gave rise to new forms.

This very important branch of science that deals with active and passive migration was founded by Darwin, with the aid of the theory of evolution; and at the same time he advanced the true explanation of the remarkable relation or similarity of the living population in any locality to the fossil forms found in it. Moritz Wagner very ably developed his idea under the title of "the theory of migration." In my opinion, this famous traveller has rather over–estimated the value of his theory of migration when he takes it to be an indispensable condition of the formation of new species and opposes the theory of selection. The two theories are not opposed in their main features. Migration (by which the stem–form of a new species is isolated) is really only a special case of selection. The striking and interesting facts of chorology can be explained only by the theory of evolution, and therefore we must count them among the most important of its inductive bases.

The same must be said of all the remarkable phenomena which we perceive in the economy of the living organism. The many and various relations of plants and animals to each other and to their environment, which are treated in bionomy (from nomos, law or norm, and bios, life), the interesting facts of parasitism, domesticity, care of the young, social habits, etc., can only be explained by the action of heredity and adaptation. Formerly people saw only the guidance of a beneficent Providence in these phenomena; to–day we discover in them admirable proofs of the theory of evolution. It is impossible to understand them except in the light of this theory and the struggle for life.

Finally, we must, in my opinion, count among the chief inductive bases of the theory of evolution the foetal development of the individual organism, the whole science of embryology or ontogeny. But as the later chapters will deal with this in detail, I need say nothing further here. I shall endeavour in the following pages to show, step by step, how the whole of the embryonic phenomena form a massive chain of proof for the theory of evolution; for they can be explained in no other way. In thus appealing to the close causal connection between ontogenesis and phylogenesis, and taking our stand throughout on the biogenetic law, we shall be able to prove, stage by stage, from the facts of embryology, the evolution of man from the lower animals.

The general adoption of the theory of evolution has definitely closed the controversy as to the nature or definition of the species. The word has no ABSOLUTE meaning whatever, but is only a group–name, or category of classification, with a purely relative value. In 1857, it is true, a famous and gifted, but inaccurate and dogmatic, scientist, Louis Agassiz, attempted to give an absolute value to these "categories of classification." He did this in his Essay on Classification, in which he turns upside down the phenomena of organic nature, and, instead of tracing them to their natural causes, examines them through a theological prism. The true species (bona species) was, he said, an "incarnate idea of the Creator." Unfortunately, this pretty phrase has no more scientific value than all the other attempts to save the absolute or intrinsic value of the species.

The dogma of the fixity and creation of species lost its last great champion when Agassiz died in 1873. The opposite theory, that all the different species descend from common stem–forms, encounters no serious difficulty to–day. All the endless research into the nature of the species, and the possibility of several species descending from a common ancestor, has been closed to–day by the removal of the sharp limits that had been set up between species and varieties on the one hand, and species and genera on the other. I gave an analytic proof of this in my monograph on the sponges (1872), having made a very close study of variability in this small but highly instructive group, and shown the impossibility of making any dogmatic distinction of species. According as the classifier takes his ideas of genus, species, and variety in a broader or in a narrower sense, he will find in the small group of the sponges either one genus with three species, or three genera with 238 species, or 113 genera with 591 species. Moreover, all these forms are so connected by intermediate forms that we can convincingly prove the descent of all the sponges from a common stem–form, the olynthus.

Here, I think, I have given an analytic solution of the problem of the origin of species, and so met the demand of certain opponents of evolution for an actual instance of descent from a stem–form. Those who are not satisfied with the synthetic proofs of the theory of evolution which are provided by comparative anatomy, embryology, paleontology, dysteleology, chorology, and classification, may try to refute the analytic proof given in my treatise on the sponge, the outcome of five years of assiduous study. I repeat: It is now impossible to oppose evolution on the ground that we have no convincing example of the descent of all the species of a group from a common ancestor. The monograph on the sponges furnishes such a proof, and, in my opinion, an indisputable proof. Any man of science who will follow the protracted steps of my inquiry and test my assertions will find that in the case of the sponges we can follow the actual evolution of species in a concrete case. And if this is so, if we can show the origin of all the species from a common form in one single class, we have the solution of the problem of man's origin, because we are in a position to prove clearly his descent from the lower animals.

At the same time, we can now reply to the often–repeated assertion, even heard from scientists of our own day, that the descent of man from the lower animals, and proximately from the apes, still needs to be "proved with certainty." These "certain proofs" have been available for a long time; one has only to open one's eyes to see them. It is a mistake to seek them in the discovery of intermediate forms between man and the ape, or the conversion of an ape into a human being by skilful education. The proofs lie in the great mass of empirical material we have already collected. They are furnished in the strongest form by the data of comparative anatomy and embryology, completed by paleontology. It is not a question now of detecting new proofs of the evolution of man, but of examining and understanding the proofs we already have.

I was almost alone thirty–six years ago when I made the first attempt, in my General Morphology, to put organic science on a mechanical foundation through Darwin's theory of descent. The association of ontogeny and phylogeny and the proof of the intimate causal connection between these two sections of the science of evolution, which I expounded in my work, met with the most spirited opposition on nearly all sides. The next ten years were a terrible "struggle for life" for the new theory. But for the last twenty–five years the tables have been turned. The phylogenetic method has met with so general a reception, and found so prolific a use in every branch of biology, that it seems superfluous to treat any further here of its validity and results. The proof of it lies in the whole morphological literature of the last three decades. But no other science has been so profoundly modified in its leading thoughts by this adoption, and been forced to yield such far–reaching consequences, as that science which I am now seeking to establish—monistic anthropogeny.

This statement may seem to be rather audacious, since the very next branch of biology, anthropology in the stricter sense, makes very little use of these results of anthropogeny, and sometimes expressly opposes them.* (*This does not apply to English anthropologists, who are almost all evolutionists.) This applies especially to the attitude which has characterised the German Anthropological Society (the Deutsche Gesellschaft fur Anthropologie) for some thirty years. Its powerful president, the famous pathologist, Rudolph Virchow, is chiefly responsible for this. Until his death (September 5th, 1902) he never ceased to reject the theory of descent as unproven, and to ridicule its chief consequence—the descent of man from a series of mammal ancestors—as a fantastic dream. I need only recall his well–known expression at the Anthropological Congress at Vienna in 1894, that "it would be just as well to say man came from the sheep or the elephant as from the ape."

Virchow's assistant, the secretary of the German Anthropological Society, Professor Johannes Ranke of Munich, has also indefatigably opposed transformism: he has succeeded in writing a work in two volumes (Der Mensch), in which all the facts relating to his organisation are explained in a sense hostile to evolution. This work has had a wide circulation, owing to its admirable illustrations and its able treatment of the most interesting facts of anatomy and physiology—exclusive of the sexual organs! But, as it has done a great deal to spread erroneous views among the general public, I have included a criticism of it in my History of Creation, as well as met Virchow's attacks on anthropogeny.

Neither Virchow, nor Ranke, nor any other "exact" anthropologist, has attempted to give any other natural explanation of the origin of man. They have either set completely aside this "question of questions" as a transcendental problem, or they have appealed to religion for its solution. We have to show that this rejection of the rational explanation is totally without justification. The fund of knowledge which has accumulated in the progress of biology in the nineteenth century is quite adequate to furnish a rational explanation, and to establish the theory of the evolution of man on the solid facts of his embryology.

CHAPTER 1.6. THE OVUM AND THE AMOEBA.

In order to understand clearly the course of human embryology, we must select the more important of its wonderful and manifold processes for fuller explanation, and then proceed from these to the innumerable features of less importance. The most important feature in this sense, and the best starting–point for ontogenetic study, is the fact that man is developed from an ovum, and that this ovum is a simple cell. The human ovum does not materially differ in form and composition from that of the other mammals, whereas there is a distinct difference between the fertilised ovum of the mammal and that of any other animal.

(FIGURE 1.1. The human ovum, magnified 100 times. The globular mass of yelk (b) is enclosed by a transparent membrane (the ovolemma or zona pellucida [a]), and contains a noncentral nucleus (the germinal vesicle, c). Cf. Figure 1.14.)

This fact is so important that few should be unaware of its extreme significance; yet it was quite unknown in the first quarter of the nineteenth century. As we have seen, the human and mammal ovum was not discovered until 1827, when Carl Ernst von Baer detected it. Up to that time the larger vesicles, in which the real and much smaller ovum is contained, had been wrongly regarded as ova. The important circumstance that this mammal ovum is a simple cell, like the ovum of other animals, could not, of course, be recognised until the cell theory was established. This was not done, by Schleiden for the plant and Schwann for the animal, until 1838. As we have seen, this cell theory is of the greatest service in explaining the human frame and its embryonic development. Hence we must say a few words about the actual condition of the theory and the significance of the views it has suggested.

In order properly to appreciate the cellular theory, the most important element in our science, it is necessary to understand in the first place that the cell is a UNIFIED

ORGANISM, a self–contained living being. When we anatomically dissect the fully–formed animal or plant into its various organs, and then examine the finer structure of these organs with the microscope, we are surprised to find that all these different parts are ultimately made up of the same structural element or unit. This common unit of structure is the cell. It does not matter whether we thus dissect a leaf, flower, or fruit, or a bone, muscle, gland, or bit of skin, etc.; we find in every case the same ultimate constituent, which has been called the cell since Schleiden's discovery. There are many opinions as to its real nature, but the essential point in our view of the cell is to look upon it as a self–contained or independent living unit. It is, in the words of Brucke, "an elementary organism." We may define it most precisely as the ultimate organic unit, and, as the cells are the sole active principles in every vital function, we may call them the "plastids," or "formative elements." This unity is found in both the anatomic structure and the physiological function. In the case of the protists, the entire organism usually consists of a single independent cell throughout life. But in the tissue–forming animals and plants, which are the great majority, the organism begins its career as a simple cell, and then grows into a cell–community, or, more correctly, an organised cell–state. Our own body is not really the simple unity that it is generally supposed to be. On the contrary, it is a very elaborate social system of countless microscopic organisms, a colony or commonwealth, made up of innumerable independent units, or very different tissue–cells.

In reality, the term "cell," which existed long before the cell theory was formulated, is not happily chosen. Schleiden, who first brought it into scientific use in the sense of the cell theory, gave this name to the elementary organisms because, when you find them in the dissected plant, they generally have the appearance of chambers, like the cells in a bee–hive, with firm walls and a fluid or pulpy content. But some cells, especially young ones, are entirely without the enveloping membrane, or stiff wall. Hence we now generally describe the cell as a living, viscous particle of protoplasm, enclosing a firmer nucleus in its albuminoid body. There may be an enclosing membrane, as there actually is in the case of most of the plants; but it may be wholly lacking, as is the case with most of the animals. There is no membrane at all in the first stage. The young cells are usually round, but they vary much in shape later on. Illustrations of this will be found in the cells of the various parts of the body shown in Figures 1.3 to 1.7.

Hence the essential point in the modern idea of the cell is that it is made up of two different active constituents—an inner and an outer part. The smaller and inner part is the nucleus (or caryon or cytoblastus, Figure 1.1 c and Figure 1.2 k). The outer and larger part, which encloses the other, is the body of the cell (celleus, cytos, or cytosoma). The soft living substance of which the two are composed has a peculiar chemical composition, and belongs to the group of the albuminoid plasma–substances ("formative matter"), or protoplasm. The essential and indispensable element of the nucleus is called nuclein (or caryoplasm); that of the cell body is called plastin (or cytoplasm). In the most rudimentary cases both substances seem to be quite simple and homogeneous, without any visible structure. But, as a rule, when we examine them under a high power of the microscope, we find a certain structure in the protoplasm. The chief and most common form of this is the fibrous or net–like "thready structure" (Frommann) and the frothy "honeycomb structure" (Butschli).

(FIGURE 1.2. Stem–cell of one of the echinoderms (cytula, or "first segmentation–cell" = fertilised ovum), after Hertwig. k is the nucleus or caryon.)

The shape or outer form of the cell is infinitely varied, in accordance with its endless power of adapting itself to the most diverse activities or environments. In its simplest form the cell is globular (Figure 1.2). This normal round form is especially found in cells of the simplest construction, and those that are developed in a free fluid without any external pressure. In such cases the nucleus also is not infrequently round, and located in the centre of the cell–body (Figure 1.2 k). In other cases, the cells have no definite shape; they are constantly changing their form owing to their automatic movements. This is the case with the amoebae (Figures 1.15 and 1.16) and the amoeboid travelling cells (Figure 1.11), and also with very young ova (Figure 1.13). However, as a rule, the cell assumes a definite form in the course of its career. In the tissues of the multicellular organism, in which a number of similar cells are bound together in virtue of certain laws of heredity, the shape is determined partly by the form of their connection and partly by their special functions. Thus, for instance, we find in the mucous lining of our tongue very thin and delicate flat cells of roundish shape (Figure 1.3). In the outer skin we find similar, but harder, covering cells, joined together by saw–like edges (Figure 1.4). In the liver and other glands there are thicker and softer cells, linked together in rows (Figure 1.5).

The last–named tissues (Figures 1.3 to 1.5) belong to the simplest and most primitive type, the group of the "covering–tissues," or epithelia. In these "primary tissues" (to which the germinal layers belong) simple cells of the same kind are arranged in layers. The arrangement and shape are more complicated in the "secondary tissues," which are gradually developed out of the primary, as in the tissues of the muscles, nerves, bones, etc. In the bones, for instance, which belong to the group of supporting or connecting organs, the cells (Figure 1.6) are star–shaped, and are joined together by numbers of net–like interlacing processes; so, also, in the tissues of the teeth (Figure 1.7), and in other forms of supporting–tissue, in which a soft or hard substance (intercellular matter, or base) is inserted between the cells.

(FIGURE 1.3. Three epithelial cells from the mucous lining of the tongue.

FIGURE 1.4. Five spiny or grooved cells, with edges joined, from the outer skin (epidermis): one of them (b) is isolated.

FIGURE 1.5. Ten liver–cells: one of them (b) has two nuclei.)

The cells also differ very much in size. The great majority of them are invisible to the naked eye, and can be seen only through the microscope (being as a rule between 1/2500 and 1/250 inch in diameter). There are many of the smaller plastids—such as the famous bacteria—which only come into view with a very high magnifying power. On the other hand, many cells attain a considerable size, and run occasionally to several inches in diameter, as do certain kinds of rhizopods among the unicellular protists (such as the radiolaria and thalamophora). Among the tissue–cells of the animal body many of the muscular fibres and nerve fibres are more than four inches, and sometimes more than a yard, in length. Among the largest cells are the yelk–filled ova; as, for instance, the yellow "yolk" in the hen's egg, which we shall describe later (Figure 1.15).

Cells also vary considerably in structure. In this connection we must first distinguish between the active and passive components of the cell. It is only the former, or active parts of the cell, that really live, and effect that marvellous world of phenomena to which we give the name of "organic life." The first of these is the inner nucleus (caryoplasm), and the second the body of the cell (cytoplasm). The passive portions come third; these are subsequently formed from the others, and I have given them the name of "plasma–products." They are partly external (cell–membranes and intercellular matter) and partly internal (cell–sap and cell–contents).

The nucleus (or caryon), which is usually of a simple roundish form, is quite structureless at first (especially in very young cells), and composed of homogeneous nuclear matter or caryoplasm (Figure 1.2 k). But, as a rule, it forms a sort of vesicle later on, in which we can distinguish a more solid nuclear base (caryobasis) and a softer or fluid nuclear sap (caryolymph). In a mesh of the nuclear network (or it may be on the inner side of the nuclear envelope) there is, as a rule, a dark, very opaque, solid body, called the nucleolus. Many of the nuclei contain several of these nucleoli (as, for instance, the germinal vesicle of the ova of fishes and amphibia). Recently a very small, but particularly important, part of the nucleus has been distinguished as the central body (centrosoma)—a tiny particle that is originally found in the nucleus itself, but is usually outside it, in the cytoplasm; as a rule, fine threads stream out from it in the cytoplasm. From the position of the central body with regard to the other parts it seems probable that it has a high physiological importance as a centre of movement; but it is lacking in many cells.

The cell–body also consists originally, and in its simplest form, of a homogeneous viscid plasmic matter. But, as a rule, only the smaller part of it is formed of the living active cell–substance (protoplasm); the greater part consists of dead, passive plasma–products (metaplasm). It is useful to distinguish between the inner and outer of these. External plasma–products (which are thrust out from the protoplasm as solid "structural matter") are the cell–membranes and the intercellular matter. The internal plasma–products are either the fluid cell–sap or hard structures. As a rule, in mature and differentiated cells these various parts are so arranged that the protoplasm (like the caryoplasm in the round nucleus) forms a sort of skeleton or framework. The spaces of this network are filled partly with the fluid cell–sap and partly by hard structural products.

(FIGURE 1.6. Nine star–shaped bone–cells, with interlaced branches.

FIGURE 1.7. Eleven star–shaped cells from the enamel of a tooth, joined together by their branchlets.)

The simple round ovum, which we take as the starting–point of our study (Figures 1.1 and 1.2), has in many cases the vague, indifferent features of the typical primitive cell. As a contrast to it, and as an instance of a very highly differentiated plastid, we may consider for a moment a large nerve–cell, or ganglionic cell, from the brain. The ovum stands potentially for the entire organism—in other words, it has the faculty of building up out of itself the whole multicellular body. It is the common parent of all the countless generations of cells which form the different tissues of the body; it unites all their powers in itself, though only potentially or in germ. In complete contrast to this, the neural cell in the brain (Figure 1.9) develops along one rigid line. It cannot, like the ovum, beget endless generations of cells, of which some will become skin–cells, others muscle–cells, and others again bone–cells. But, on the other hand, the nerve–cell has become fitted to discharge the highest functions of life; it has the powers of sensation, will, and thought. It is a real soul–cell, or an elementary organ of the psychic activity. It has, therefore, a most elaborate and delicate structure. Numbers of extremely fine threads, like the electric wires at a large telegraphic centre, cross and recross in the delicate protoplasm of the nerve cell, and pass out in the branching processes which proceed from it and put it in communication with other nerve–cells or nerve–fibres (a, b). We can only partly follow their intricate paths in the fine matter of the body of the cell.

Here we have a most elaborate apparatus, the delicate structure of which we are just beginning to appreciate through our most powerful microscopes, but whose significance is rather a matter of conjecture than knowledge. Its intricate structure corresponds to the very complicated functions of the mind. Nevertheless, this elementary organ of psychic activity—of which there are thousands in our brain—is nothing but a single cell. Our whole mental life is only the joint result of the combined activity of all these nerve–cells, or soul–cells. In the centre of each cell there is a large transparent nucleus, containing a small and dark nuclear body. Here, as elsewhere, it is the nucleus that determines the individuality of the cell; it proves that the whole structure, in spite of its intricate composition, amounts to only a single cell.

(FIGURE 1.8. Unfertilised ovum of an echinoderm (from Hertwig). The vesicular nucleus (or "germinal vesicle") is globular, half the size of the round ovum, and encloses a nuclear framework, in the central knot of which there is a dark nucleolus (the "germinal spot").

FIGURE 1.9. A large branching nerve–cell, or "soul–cell," from the brain of an electric fish (Torpedo), magnified 600 times. In the middle of the cell is the large transparent round nucleus, one nucleolus, and, within the latter again, a nucleolinus. The protoplasm of the cell is split into innumerable fine threads (or fibrils), which are embedded in intercellular matter, and are prolonged into the branching processes of the cell (b). One branch (a) passes into a nerve–fibre. (From Max Schultze.))

In contrast with this very elaborate and very strictly differentiated psychic cell (Figure 1.9), we have our ovum (Figures 1.1 and 1.2), which has hardly any structure at all. But even in the case of the ovum we must infer from its properties that its protoplasmic body has a very complicated chemical composition and a fine molecular structure which escapes our observation. This presumed molecular structure of the plasm is now generally admitted; but it has never been seen, and, indeed, lies far beyond the range of microscopic vision. It must not be confused—as is often done—with the structure of the plasm (the fibrous network, groups of granules, honey–comb, etc.) which does come within the range of the microscope.

But when we speak of the cells as the elementary organisms, or structural units, or "ultimate individualities," we must bear in mind a certain restriction of the phrases. I mean, that the cells are not, as is often supposed, the very lowest stage of organic individuality. There are yet more elementary organisms to which I must refer occasionally. These are what we call the "cytodes" (cytos = cell), certain living, independent beings, consisting only of a particle of plasson—an albuminoid substance, which is not yet differentiated into caryoplasm and cytoplasm, but combines the properties of both. Those remarkable beings called the monera—especially the chromacea and bacteria—are specimens of these simple cytodes. (Compare Chapter 2.19.) To be quite accurate, then, we must say: the elementary organism, or the ultimate individual, is found in two different stages. The first and lower stage is the cytode, which consists merely of a particle of plasson, or quite simple plasm. The second and higher stage is the cell, which is already divided or differentiated into nuclear matter and cellular matter. We comprise both kinds—the cytodes and the cells—under the name of plastids ("formative particles"), because they are the real builders of the organism. However, these cytodes are not found, as a rule, in the higher animals and plants; here we have only real cells with a nucleus. Hence, in these tissue–forming organisms (both plant and animal) the organic unit always consists of two chemically and anatomically different parts—the outer cell–body and the inner nucleus.

In order to convince oneself that this cell is really an independent organism, we have only to observe the development and vital phenomena of one of them. We see then that it performs all the essential functions of life—both vegetal and animal—which we find in the entire organism. Each of these tiny beings grows and nourishes itself independently. It takes its food from the surrounding fluid; sometimes, even, the naked cells take in solid particles at certain points of their surface—in other words, "eat" them—without needing any special mouth and stomach for the purpose (cf. Figure 1.19).

Further, each cell is able to reproduce itself. This multiplication, in most cases, takes the form of a simple cleavage, sometimes direct, sometimes indirect; the simple direct (or "amitotic") division is less common, and is found, for instance, in the blood cells (Figure 1.10). In these the nucleus first divides into two equal parts by constriction. The indirect (or "mitotic") cleavage is much more frequent; in this the caryoplasm of the nucleus and the cytoplasm of the cell–body act upon each other in a peculiar way, with a partial dissolution (caryolysis), the formation of knots and loops (mitosis), and a movement of the halved plasma–particles towards two mutually repulsive poles of attraction (caryokinesis, Figure 1.11.)

(FIGURE 1.10. Blood–cells, multiplying by direct division, from the blood of the embryo of a stag. Originally, each blood–cell has a nucleus and is round (a). When it is going to multiply, the nucleus divides into two (b, c, d). Then the protoplasmic body is constricted between the two nuclei, and these move away from each other (e). Finally, the constriction is complete, and the cell splits into two daughter–cells (f). (From Frey.))

FIGURE 1.11. Indirect or mitotic cell–division (with caryolysis and caryokinesis) from the skin of the larva of a salamander. (From Rabl.).
  A. Mother–cell (Knot, spirema), with Nuclear threads (chromosomata)
    (coloured nuclear matter, chromatin), Cytosoma, Nuclear membrane,
    Protoplasm of the cell–body and Nuclear sap. B. Mother–star, the loops beginning to split lengthways (nuclear membrane gone), with Star–like appearance in cytoplasm, Centrosoma (sphere of attraction), Nuclear spindle (achromin, colourless matter) and Nuclear loops (chromatin, coloured matter). C. The two daughter–stars, produced by the breaking of the loops of the mother–star (moving away), with Upper daughter–crown, Connecting threads of the two crowns (achromin), Lower daughter–crown and Double–star (amphiaster). D. The two daughter–cells, produced by the complete division of the two nuclear halves (cytosomata still connected at the equator) (Double–knot, Dispirema), with Upper daughter–nucleus, Equatorial constriction of the cell–body and Lower daughter–nucleus.)

The intricate physiological processes which accompany this "mitosis" have been very closely studied of late years. The inquiry has led to the detection of certain laws of evolution which are of extreme importance in connection with heredity. As a rule, two very different parts of the nucleus play an important part in these changes. They are: the chromatin, or coloured nuclear substance, which has a peculiar property of tingeing itself deeply with certain colouring matters (carmine, haematoxylin, etc.), and the achromin (or linin, or achromatin), a colourless nuclear substance that lacks this property. The latter generally forms in the dividing cell a sort of spindle, at the poles of which there is a very small particle, also colourless, called the "central body" (centrosoma). This acts as the centre or focus in a "sphere of attraction" for the granules of protoplasm in the surrounding cell–body, and assumes a star–like appearance (the cell–star, or monaster). The two central bodies, standing opposed to each other at the poles of the nuclear spindle, form "the double–star" (or amphiaster, Figure 1.11, BC). The chromatin often forms a long, irregularly–wound thread—"the coil" (spirema, Figure A). At the commencement of the cleavage it gathers at the equator of the cell, between the stellar poles, and forms a crown of U–shaped loops (generally four or eight, or some other definite number). The loops split lengthwise into two halves (B), and these back away from each other towards the poles of the spindle (C). Here each group forms a crown once more, and this, with the corresponding half of the divided spindle, forms a fresh nucleus (D). Then the protoplasm of the cell–body begins to contract in the middle, and gather about the new daughter–nuclei, and at last the two daughter–cells become independent beings.

Between this common mitosis, or indirect cell–division—which is the normal cleavage–process in most cells of the higher animals and plants—and the simple direct division (Figure 1.10) we find every grade of segmentation; in some circumstances even one kind of division may be converted into another.

The plastid is also endowed with the functions of movement and sensation. The single cell can move and creep about, when it has space for free movement and is not prevented by a hard envelope; it then thrusts out at its surface processes like fingers, and quickly withdraws them again, and thus changes its shape (Figure 1.12). Finally, the young cell is sensitive, or more or less responsive to stimuli; it makes certain movements on the application of chemical and mechanical irritation. Hence we can ascribe to the individual cell all the chief functions which we comprehend under the general heading of "life"—sensation, movement, nutrition, and reproduction. All these properties of the multicellular and highly developed animal are also found in the single animal–cell, at least in its younger stages. There is no longer any doubt about this, and so we may regard it as a solid and important base of our physiological conception of the elementary organism.

Without going any further here into these very interesting phenomena of the life of the cell, we will pass on to consider the application of the cell theory to the ovum. Here comparative research yields the important result that EVERY OVUM IS AT FIRST A SIMPLE CELL. I say this is very important, because our whole science of embryology now resolves itself into the problem: "How does the multicellular organism arise from the unicellular?" Every organic individual is at first a simple cell, and as such an elementary organism, or a unit of individuality. This cell produces a cluster of cells by segmentation, and from these develops the multicellular organism, or individual of higher rank.

When we examine a little closer the original features of the ovum, we notice the extremely significant fact that in its first stage the ovum is just the same simple and indefinite structure in the case of man and all the animals (Figure 1.13). We are unable to detect any material difference between them, either in outer shape or internal constitution. Later, though the ova remain unicellular, they differ in size and shape, enclose various kinds of yelk–particles, have different envelopes, and so on. But when we examine them at their birth, in the ovary of the female animal, we find them to be always of the same form in the first stages of their life. In the beginning each ovum is a very simple, roundish, naked, mobile cell, without a membrane; it consists merely of a particle of cytoplasm enclosing a nucleus (Figure 1.13). Special names have been given to these parts of the ovum; the cell–body is called the yelk (vitellus), and the cell–nucleus the germinal vesicle. As a rule, the nucleus of the ovum is soft, and looks like a small pimple or vesicle. Inside it, as in many other cells, there is a nuclear skeleton or frame and a third, hard nuclear body (the nucleolus). In the ovum this is called the germinal spot. Finally, we find in many ova (but not in all) a still further point within the germinal spot, a "nucleolin," which goes by the name of the germinal point. The latter parts (germinal spot and germinal point) have, apparently, a minor importance, in comparison with the other two (the yelk and germinal vesicle). In the yelk we must distinguish the active formative yelk (or protoplasm = first plasm) from the passive nutritive yelk (or deutoplasm = second plasm).

(FIGURE 1.12. Mobile cells from the inflamed eye of a frog (from the watery fluid of the eye, the humor aqueus). The naked cells creep freely about, by (like the amoeba or rhizopods) protruding fine processes from the uncovered protoplasmic body. These bodies vary continually in number, shape, and size. The nucleus of these amoeboid lymph–cells ("travelling cells," or planocytes) is invisible, because concealed by the numbers of fine granules which are scattered in the protoplasm. (From Frey.))

In many of the lower animals (such as sponges, polyps, and medusae) the naked ova retain their original simple appearance until impregnation. But in most animals they at once begin to change; the change consists partly in the formation of connections with the yelk, which serve to nourish the ovum, and partly of external membranes for their protection (the ovolemma, or prochorion). A membrane of this sort is formed in all the mammals in the course of the embryonic process. The little globule is surrounded by a thick capsule of glass−like transparency, the zona pellucida, or ovolemma pellucidum (Figure 1.14). When we examine it closely under the microscope, we see very fine radial streaks in it, piercing the zona, which are really very narrow canals. The human ovum, whether fertilised or not, cannot be distinguished from that of most of the other mammals. It is nearly the same everywhere in form, size, and composition. When it is fully formed, it has a diameter of (on an average) about 1/120 of an inch. When the mammal ovum has been carefully isolated, and held against the light on a glass−plate, it may be seen as a fine point even with the naked eye. The ova of most of the higher mammals are about the same size. The diameter of the ovum is almost always between 1/250 to 1/125 inch. It has always the same globular shape; the same characteristic membrane; the same transparent germinal vesicle with its dark germinal spot. Even when we use the most powerful microscope with its highest power, we can detect no material difference between the ova of man, the ape, the dog, and so on. I do not mean to say that there are no differences between the ova of these different mammals. On the contrary, we are bound to assume that there are such, at least as regards chemical composition. Even the ova of different men must differ from each other; otherwise we should not have a different individual from each ovum. It is true that our crude and imperfect apparatus cannot detect these subtle individual differences, which are probably in the molecular structure. However, such a striking resemblance of their ova in form, so great as to seem to be a complete similarity, is a strong proof of the common parentage of man and the other mammals. From the common germ−form we infer a common stem−form. On the other hand, there are striking peculiarities by which we can easily distinguish the fertilised ovum of the mammal from the fertilised ovum of the birds, amphibia, fishes, and other vertebrates (see the close of Chapter 2.29).

(FIGURE 1.13. Ova of various animals, executing amoeboid movements, highly magnified. All the ova are naked cells of varying shape. In the dark fine−grained protoplasm (yelk) is a large vesicular nucleus (the germinal vesicle), and in this is seen a nuclear body (the germinal spot), in which again we often see a germinal point. Figures A1 to A4 represent the ovum of a sponge (Leuculmis echinus) in four successive movements. B1 to B8 are the ovum of a parasitic crab (Chondracanthus cornutus), in eight successive movements. (From Edward von Beneden.) C1 to C5 show the ovum of the cat in various stages of movement (from Pfluger); Figure P the ovum of a trout; E the ovum of a chicken; F a human ovum.)

The fertilised bird–ovum (Figure 1.15) is notably different. It is true that in its earliest stage (Figure 1.13 E) this ovum also is very like that of the mammal (Figure 1.13 F). But afterwards, while still within the oviduct, it takes up a quantity of nourishment and works this into the familiar large yellow yelk. When we examine a very young ovum in the hen's oviduct, we find it to be a simple, small, naked, amoeboid cell, just like the young ova of other animals (Figure 1.13). But it then grows to the size we are familiar with in the round yelk of the egg. The nucleus of the ovum, or the germinal vesicle, is thus pressed right to the surface of the globular ovum, and is embedded there in a small quantity of transparent matter, the so-called white yelk. This forms a round white spot, which is known as the "tread" (cicatricula) (Figure 1.15 b). From the tread a thin column of the white yelk penetrates through the yellow yelk to the centre of the globular cell, where it swells into a small, central globule (wrongly called the yelk–cavity, or latebra, Figure 1.15 d apostrophe). The yellow yelk–matter which surrounds this white yelk has the appearance in the egg (when boiled hard) of concentric layers (c). The yellow yelk is also enclosed in a delicate structureless membrane (the membrana vitellina, a).

As the large yellow ovum of the bird attains a diameter of several inches in the bigger birds, and encloses round yelk–particles, there was formerly a reluctance to consider it as a simple cell. This was a mistake. Every animal that has only one cell–nucleus, every amoeba, every gregarina, every infusorium, is unicellular, and remains unicellular whatever variety of matter it feeds on. So the ovum remains a simple cell, however much yellow yelk it afterwards accumulates within its protoplasm. It is, of course, different, with the bird's egg when it has been fertilised. The ovum then consists of as many cells as there are nuclei in the tread. Hence, in the fertilised egg which we eat daily, the yellow yelk is already a multicellular body. Its tread is composed of several cells, and is now commonly called the germinal disc. We shall return to this discogastrula in Chapter 1.9.

(FIGURE 1.14. The human ovum, taken from the female ovary, magnified 500 times. The whole ovum is a simple round cell. The chief part of the globular mass is formed by the nuclear yelk (deutoplasm), which is evenly distributed in the active protoplasm, and consists of numbers of fine yelk–granules. In the upper part of the yelk is the transparent round germinal vesicle, which corresponds to the nucleus. This encloses a darker granule, the germinal spot, which shows a nucleolus. The globular yelk is surrounded by the thick transparent germinal membrane (ovolemma, or zona pellucida). This is traversed by numbers of lines as fine as hairs, which are directed radially towards the centre of the ovum. These are called the pore−canals; it is through these that the moving spermatozoa penetrate into the yelk at impregnation.

FIGURE 1.15. A fertilised ovum from the oviduct of a hen. the yellow yelk (c) consists of several concentric layers (d), and is enclosed in a thin yelk−membrane (a). The nucleus or germinal vesicle is seen above in the cicatrix or "tread" (b). From that point the white yelk penetrates to the central yelk−cavity (d apostrophe). The two kinds of yelk do not differ very much.

FIGURE 1.16. A creeping amoeba (highly magnified). The whole organism is a simple naked cell, and moves about by means of the changing arms which it thrusts out of and withdraws into its protoplasmic body. Inside it is the roundish nucleus with its nucleolus.)

When the mature bird−ovum has left the ovary and been fertilised in the oviduct, it covers itself with various membranes which are secreted from the wall of the oviduct. First, the large clear albuminous layer is deposited around the yellow yelk; afterwards, the hard external shell, with a fine inner skin. All these gradually forming envelopes and processes are of no importance in the formation of the embryo; they serve merely for the protection of the original simple ovum. We sometimes find extraordinarily large eggs with strong envelopes in the case of other animals, such as fishes of the shark type. Here, also, the ovum is originally of the same character as it is in the mammal; it is a perfectly simple and naked cell. But, as in the case of the bird, a considerable quantity of nutritive yelk is accumulated inside the original yelk as food for the developing embryo; and various coverings are formed round the egg. The ovum of many other animals has the same internal and external features. They have, however, only a physiological, not a morphological, importance; they have no direct influence on the formation of the foetus. They are partly consumed as food by the embryo, and partly serve as protective envelopes. Hence we may leave them out of consideration altogether here, and restrict ourselves to material points—TO THE SUBSTANTIAL IDENTITY OF THE ORIGINAL OVUM IN MAN AND THE REST OF THE ANIMALS (Figure 1.13).

Now, let us for the first time make use of our biogenetic law; and directly apply this fundamental law of evolution to the human ovum. We reach a very simple, but very important, conclusion. FROM THE FACT THAT THE HUMAN OVUM AND THAT OF ALL OTHER ANIMALS CONSISTS OF A SINGLE CELL, IT FOLLOWS IMMEDIATELY, ACCORDING TO THE BIOGENETIC LAW, THAT ALL THE ANIMALS, INCLUDING MAN, DESCEND FROM A UNICELLULAR ORGANISM. If our biogenetic law is true, if the embryonic development is a summary or condensed recapitulation of the stem–history—and there can be no doubt about it—we are bound to conclude, from the fact that all the ova are at first simple cells, that all the multicellular organisms originally sprang from a unicellular being. And as the original ovum in man and all the other animals has the same simple and indefinite appearance, we may assume with some probability that this unicellular stem–form was the common ancestor of the whole animal world, including man. However, this last hypothesis does not seem to me as inevitable and as absolutely certain as our first conclusion.

This inference from the unicellular embryonic form to the unicellular ancestor is so simple, but so important, that we cannot sufficiently emphasise it. We must, therefore, turn next to the question whether there are to–day any unicellular organisms, from the features of which we may draw some approximate conclusion as to the unicellular ancestors of the multicellular organisms. The answer is: Most certainly there are. There are assuredly still unicellular organisms which are, in their whole nature, really nothing more than permanent ova. There are independent unicellular organisms of the simplest character which develop no further, but reproduce themselves as such, without any further growth. We know to–day of a great number of these little beings, such as the gregarinae, flagellata, acineta, infusoria, etc. However, there is one of them that has an especial interest for us, because it at once suggests itself when we raise our question, and it must be regarded as the unicellular being that approaches nearest to the real ancestral form. This organism is the amoeba.

For a long time now we have comprised under the general name of amoebae a number of microscopic unicellular organisms, which are very widely distributed, especially in fresh–water, but also in the ocean; in fact, they have lately been discovered in damp soil. There are also parasitic amoebae which live inside other animals. When we place one of these amoebae in a drop of water under the microscope and examine it with a high power, it generally appears as a roundish particle of a very irregular and varying shape (Figures 1.16 and 1.17). In its soft, slimy, semi–fluid substance, which consists of protoplasm, we see only the solid globular particle it contains, the nucleus. This unicellular body moves about continually, creeping in every direction on the glass on which we are examining it. The movement is effected by the shapeless body thrusting out finger–like processes at various parts of its surface; and these are slowly but continually changing, and drawing the rest of the body after them. After a time, perhaps, the action changes. The amoeba suddenly stands still, withdraws its projections, and assumes a globular shape. In a little while, however, the round body begins to expand again, thrusts out arms in another direction, and moves on once more. These changeable processes are called "false feet," or pseudopodia, because they act physiologically as feet, yet are not special organs in the anatomic sense. They disappear as quickly as they come, and are nothing more than temporary projections of the semi–fluid and structureless body.

(FIGURE 1.17. Division of a unicellular amoeba (Amoeba polypodia) in six stages. (From F.E. Schultze.) the dark spot is the nucleus, the lighter spot a contractile vacuole in the protoplasm. The latter reforms in one of the daughter–cells.)

FIGURE 1.18. Ovum of a sponge (Olynthus). The ovum creeps about in a body of the sponge by thrusting out ever–changing processes. It is indistinguishable from the common amoeba.)

If you touch one of these creeping amoebae with a needle, or put a drop of acid in the water, the whole body at once contracts in consequence of this mechanical or physical stimulus. As a rule, the body then resumes its globular shape. In certain circumstances—for instance, if the impurity of the water lasts some time—the amoeba begins to develop a covering. It exudes a membrane or capsule, which immediately hardens, and assumes the appearance of a round cell with a protective membrane. The amoeba either takes its food directly by imbibition of matter floating in the water, or by pressing into its protoplasmic body solid particles with which it comes in contact. The latter process may be observed at any moment by forcing it to eat. If finely ground colouring matter, such as carmine or indigo, is put into the water, you can see the body of the amoeba pressing these coloured particles into itself, the substance of the cell closing round them. The amoeba can take in food in this way at any point on its surface, without having any special organs for intussusception and digestion, or a real mouth or gut.

The amoeba grows by thus taking in food and dissolving the particles eaten in its protoplasm. When it reaches a certain size by this continual feeding, it begins to reproduce. This is done by the simple process of cleavage (Figure 1.17). First, the nucleus divides into two parts. Then the protoplasm is separated between the two new nuclei, and the whole cell splits into two daughter–cells, the protoplasm gathering about each of the nuclei. The thin bridge of protoplasm which at first connects the daughter–cells soon breaks. Here we have the simple form of direct cleavage of the nuclei. Without mitosis, or formation of threads, the homogeneous nucleus divides into two halves. These move away from each other, and become centres of attraction for the enveloping matter, the protoplasm. The same direct cleavage of the nuclei is also witnessed in the reproduction of many other protists, while other unicellular organisms show the indirect division of the cell.

Hence, although the amoeba is nothing but a simple cell, it is evidently able to accomplish all the functions of the multicellular organism. It moves, feels, nourishes itself, and reproduces. Some kinds of these amoebae can be seen with the naked eye, but most of them are microscopically small. It is for the following reasons that we regard the amoebae as the unicellular organisms which have special phylogenetic (or evolutionary) relations to the ovum. In many of the lower animals the ovum retains its original naked form until fertilisation, develops no membranes, and is then often indistinguishable from the ordinary amoeba. Like the amoebae, these naked ova may thrust out processes, and move about as travelling cells. In the sponges these mobile ova move about freely in the maternal body like independent amoebae (Figure 1.17). They had been observed by earlier scientists, but described as foreign bodies—namely, parasitic amoebae, living parasitically on the body of the sponge. Later, however, it was discovered that they were not parasites, but the ova of the sponge. We also find this remarkable phenomenon among other animals, such as the graceful, bell–shaped zoophytes, which we call polyps and medusae. Their ova remain naked cells, which thrust out amoeboid projections, nourish themselves, and move about. When they have been fertilised, the multicellular organism is formed from them by repeated segmentation.

It is, therefore, no audacious hypothesis, but a perfectly sound conclusion, to regard the amoeba as the particular unicellular organism which offers us an approximate illustration of the ancient common unicellular ancestor of all the metazoa, or multicellular animals. The simple naked amoeba has a less definite and more original character than any other cell. Moreover, there is the fact that recent research has discovered such amoeba–like cells everywhere in the mature body of the multicellular animals. They are found, for instance, in the human blood, side by side with the red corpuscles, as colourless blood–cells; and it is the same with all the vertebrates. They are also found in many of the invertebrates—for instance, in the blood of the snail. I showed, in 1859, that these colourless blood–cells can, like the independent amoebae, take up solid particles, or "eat" (whence they are called phagocytes = "eating–cells," Figure 1.19). Lately, it has been discovered that many different cells may, if they have room enough, execute the same movements, creeping about and eating. They behave just like amoebae (Figure 1.12). It has also been shown that these "travelling–cells," or planocytes, play an important part in man's physiology and pathology (as means of transport for food, infectious matter, bacteria, etc.).

The power of the naked cell to execute these characteristic amoeba–like movements comes from the contractility (or automatic mobility) of its protoplasm. This seems to be a universal property of young cells. When they are not enclosed by a firm membrane, or confined in a "cellular prison," they can always accomplish these amoeboid movements. This is true of the naked ova as well as of any other naked cells, of the "travelling–cells," of various kinds in connective tissue, lymph–cells, mucus–cells, etc.

We have now, by our study of the ovum and the comparison of it with the amoeba, provided a perfectly sound and most valuable foundation for both the embryology and the evolution of man. We have learned that the human ovum is a simple cell, that this ovum is not materially different from that of other mammals, and that we may infer from it the existence of a primitive unicellular ancestral form, with a substantial resemblance to the amoeba.

The statement that the earliest progenitors of the human race were simple cells of this kind, and led an independent unicellular life like the amoeba, has not only been ridiculed as the dream of a natural philosopher, but also been violently censured in theological journals as "shameful and immoral." But, as I observed in my essay On the Origin and Ancestral Tree of the Human Race in 1870, this offended piety must equally protest against the "shameful and immoral" fact that each human individual is developed from a simple ovum, and that this human ovum is indistinguishable from those of the other mammals, and in its earliest stage is like a naked amoeba. We can show this to be a fact any day with the microscope, and it is little use to close one's eyes to "immoral" facts of this kind. It is as indisputable as the momentous conclusions we draw from it and as the vertebrate character of man (see Chapter 1.11).

(FIGURE 1.19. Blood–cells that eat, or phagocytes, from a naked sea–snail (Thetis), greatly magnified. I was the first to observe in the blood–cells of this snail the important fact that "the blood–cells of the invertebrates are unprotected pieces of plasm, and take in food, by means of their peculiar movements, like the amoebae." I had (in Naples, on May 10th, 1859) injected into the blood–vessels of one of these snails an infusion of water and ground indigo, and was greatly astonished to find the blood–cells themselves more or less filled with the particles of indigo after a few hours. After repeated injections I succeeded in "observing the very entrance of the coloured particles in the blood–cells, which took place just in the same way as with the amoeba." I have given further particulars about this in my Monograph on the Radiolaria.)

We now see very clearly how extremely important the cell theory has been for our whole conception of organic nature. "Man's place in nature" is settled beyond question by it. Apart from the cell theory, man is an insoluble enigma to us. Hence philosophers, and especially physiologists, should be thoroughly conversant with it. The soul of man can only be really understood in the light of the cell–soul, and we have the simplest form of this in the amoeba. Only those who are acquainted with the simple psychic functions of the unicellular organisms and their gradual evolution in the series of lower animals can understand how the elaborate mind of the higher vertebrates, and especially of man, was gradually evolved from them. The academic psychologists who lack this zoological equipment are unable to do so.

This naturalistic and realistic conception is a stumbling–block to our modern idealistic metaphysicians and their theological colleagues. Fenced about with their transcendental and dualistic prejudices, they attack not only the monistic system we establish on our scientific knowledge, but even the plainest facts which go to form its foundation. An instructive instance of this was seen a few years ago, in the academic discourse delivered by a distinguished theologian, Willibald Beyschlag, at Halle, January 12th, 1900, on the occasion of the centenary festival. The theologian protested violently against the "materialistic dustmen of the scientific world who offer our people the diploma of a descent from the ape, and would prove to them that the genius of a Shakespeare or a Goethe is merely a distillation from a drop of primitive mucus." Another well–known theologian protested against "the horrible idea that the greatest of men, Luther and Christ, were descended from a mere globule of protoplasm." Nevertheless, not a single informed and impartial scientist doubts the fact that these greatest men were, like all other men—and all other vertebrates—developed from an impregnated ovum, and that this simple nucleated globule of protoplasm has the same chemical constitution in all the mammals.

CHAPTER 1.7. CONCEPTION.

The recognition of the fact that every man begins his individual existence as a simple cell is the solid foundation of all research into the genesis of man. From this fact we are forced, in virtue of our biogenetic law, to draw the weighty phylogenetic conclusion that the earliest ancestors of the human race were also unicellular organisms; and among these protozoa we may single out the vague form of the amoeba as particularly important (cf. Chapter 1.6). That these unicellular ancestral forms did once exist follows directly from the phenomena which we perceive every day in the fertilised ovum. The development of the multicellular organism from the ovum, and the formation of the germinal layers and the tissues, follow the same laws in man and all the higher animals. It will, therefore, be our next task to consider more closely the impregnated ovum and the process of conception which produces it.

The process of impregnation or sexual conception is one of those phenomena that people love to conceal behind the mystic veil of supernatural power. We shall soon see, however, that it is a purely mechanical process, and can be reduced to familiar physiological functions. Moreover, this process of conception is of the same type, and is effected by the same organs, in man as in all the other mammals. The pairing of the male and female has in both cases for its main purpose the introduction of the ripe matter of the male seed or sperm into the female body, in the sexual canals of which it encounters the ovum. Conception then ensues by the blending of the two.

We must observe, first, that this important process is by no means so widely distributed in the animal and plant world as is commonly supposed. There is a very large number of lower organisms which propagate unsexually, or by monogamy; these are especially the sexless monera (chromacea, bacteria, etc.) but also many other protists, such as the amoebae, foraminifera, radiolaria, myxomycetae, etc. In these the multiplication of individuals takes place by unsexual reproduction, which takes the form of cleavage, budding, or spore−formation. The copulation of two coalescing cells, which in these cases often precedes the reproduction, cannot be regarded as a sexual act unless the two copulating plastids differ in size or structure. On the other hand, sexual reproduction is the general rule with all the higher organisms, both animal and plant; very rarely do we find asexual reproduction among them. There are, in particular, no cases of parthenogenesis (virginal conception) among the vertebrates.

Sexual reproduction offers an infinite variety of interesting forms in the different classes of animals and plants, especially as regards the mode of conception, and the conveyance of the spermatozoon to the ovum. These features are of great importance not only as regards conception itself, but for the development of the organic form, and especially for the differentiation of the sexes. There is a particularly curious correlation of plants and animals in this respect. The splendid studies of Charles Darwin and Hermann Muller on the fertilisation of flowers by insects have given us very interesting particulars of this.* (* See Darwin's work, On the Various Contrivances by which Orchids are Fertilised (1862).) This reciprocal service has given rise to a most intricate sexual apparatus. Equally elaborate structures have been developed in man and the higher animals, serving partly for the isolation of the sexual products on each side, partly for bringing them together in conception. But, however interesting these phenomena are in themselves, we cannot go into them here, as they have only a minor importance—if any at all—in the real process of conception. We must, however, try to get a very clear idea of this process and the meaning of sexual reproduction.

In every act of conception we have, as I said, to consider two different kinds of cells—a female and a male cell. The female cell of the animal organism is always called the ovum (or ovulum, egg, or egg–cell); the male cells are known as the sperm or seed–cells, or the spermatozoa (also spermium and zoospermium). The ripe ovum is, on the whole, one of the largest cells we know. It attains colossal dimensions when it absorbs great quantities of nutritive yelk, as is the case with birds and reptiles and many of the fishes. In the great majority of the animals the ripe ovum is rich in yelk and much larger than the other cells. On the other hand, the next cell which we have to consider in the process of conception, the male sperm–cell or spermatozoon, is one of the smallest cells in the animal body. Conception usually consists in the bringing into contact with the ovum of a slimy fluid secreted by the male, and this may take place either inside or out of the female body. This fluid is called sperm, or the male seed. Sperm, like saliva or blood, is not a simple fluid, but a thick agglomeration of innumerable cells, swimming about in a comparatively small quantity of fluid. It is not the fluid, but the independent male cells that swim in it, that cause conception.

(FIGURE 1.20. Spermia or spermatozoa of various mammals. The pear–shaped flattened nucleus is seen from the front in I and sideways in II. k is the nucleus, m its middle part (protoplasm), s the mobile, serpent–like tail (or whip); M four human spermatozoa, A spermatozoa from the ape; K from the rabbit; H from the mouse; C from the dog; S from the pig.

FIGURE 1.21. Spermatozoa or spermidia of various animals. (From Lang). a of a fish, b of a turbellaria worm (with two side–lashes), c to e of a nematode worm (amoeboid spermatozoa), f from a craw fish (star–shaped), g from the salamander (with undulating membrane), h of an annelid (a and h are the usual shape).

FIGURE 1.22. A single human spermatozoon magnified 2000 times; a shows it from the broader and b from the narrower side. k head (with nucleus), m middle–stem, h long–stem, and e tail. (From Retzius.))

The spermatozoa of the great majority of animals have two characteristic features. Firstly, they are extraordinarily small, being usually the smallest cells in the body; and, secondly, they have, as a rule, a peculiarly lively motion, which is known as spermatozoic motion. The shape of the cell has a good deal to do with this motion. In most of the animals, and also in many of the lower plants (but not the higher) each of these spermatozoa has a very small, naked cell–body, enclosing an elongated nucleus, and a long thread hanging from it (Figure 1.20). It was long before we could recognise that these structures are simple cells. They were formerly held to be special organisms, and were called "seed animals" (spermato–zoa, or spermato–zoidia); they are now scientifically known as spermia or spermidia, or as spermatosomata (seed–bodies) or spermatofila (seed threads). It took a good deal of comparative research to convince us that each of these spermatozoa is really a simple cell. They have the same shape as in many other vertebrates and most of the invertebrates. However, in many of the lower animals they have quite a different shape. Thus, for instance, in the craw fish they are large round cells, without any movement, equipped with stiff outgrowths like bristles (Figure 1.21 f). They have also a peculiar form in some of the worms, such as the thread–worms (filaria); in this case they are sometimes amoeboid and like very small ova (Figure 1.21 c to e). But in most of the lower animals (such as the sponges and polyps) they have the same pine–cone shape as in man and the other animals (Figure 1.21 a, h).

When the Dutch naturalist Leeuwenhoek discovered these thread–like lively particles in 1677 in the male sperm, it was generally believed that they were special, independent, tiny animalcules, like the infusoria, and that the whole mature organism existed already, with all its parts, but very small and packed together, in each spermatozoon (see Chapter 1.2). We now know that the mobile spermatozoa are nothing but simple and real cells, of the kind that we call "ciliated" (equipped with lashes, or cilia). In the previous illustrations we have distinguished in the spermatozoon a head, trunk, and tail. The "head" (Figure 1.20 k) is merely the oval nucleus of the cell; the body or middle–part (m) is an accumulation of cell–matter; and the tail (s) is a thread–like prolongation of the same.

Moreover, we now know that these spermatozoa are not at all a peculiar form of cell; precisely similar cells are found in various other parts of the body. If they have many short threads projecting, they are called ciliated; if only one long, whip–shaped process (or, more rarely, two or four), caudate (tailed) cells.

Very careful recent examination of the spermia, under a very high microscopic power (Figure 1.22 a, b), has detected some further details in the finer structure of the ciliated cell, and these are common to man and the anthropoid ape. The head (k) encloses the elliptic nucleus in a thin envelope of cytoplasm; it is a little flattened on one side, and thus looks rather pear–shaped from the front (b). In the central piece (m) we can distinguish a short neck and a longer connective piece (with central body). The tail consists of a long main section (h) and a short, very fine tail (e).

The process of fertilisation by sexual conception consists, therefore, essentially in the coalescence and fusing together of two different cells. The lively spermatozoon travels towards the ovum by its serpentine movements, and bores its way into the female cell (Figure 1.23). The nuclei of both sexual cells, attracted by a certain "affinity," approach each other and melt into one.

The fertilised cell is quite another thing from the unfertilised cell. For if we must regard the spermia as real cells no less than the ova, and the process of conception as a coalescence of the two, we must consider the resultant cell as a quite new and independent organism. It bears in the cell and nuclear matter of the penetrating spermatozoon a part of the father's body, and in the protoplasm and caryoplasm of the ovum a part of the mother's body. This is clear from the fact that the child inherits many features from both parents. It inherits from the father by means of the spermatozoon, and from the mother by means of the ovum. The actual blending of the two cells produces a third cell, which is the germ of the child, or the new organism conceived. One may also say of this sexual coalescence that the STEM–CELL IS A SIMPLE HERMAPHRODITE; it unites both sexual substances in itself.

(FIGURE 1.23. The fertilisation of the ovum by the spermatozoon (of a mammal). One of the many thread–like, lively spermidia pierces through a fine pore–canal into the nuclear yelk. The nucleus of the ovum is invisible.

FIGURE 1.24. An impregnated echinoderm ovum, with small homogeneous nucleus (e k). (From Hertwig.))

I think it necessary to emphasise the fundamental importance of this simple, but often unappreciated, feature in order to have a correct and clear idea of conception. With that end, I have given a special name to the new cell from which the child develops, and which is generally loosely called "the fertilised ovum," or "the first segmentation sphere." I call it "the stem–cell" (cytula). The name "stem–cell" seems to me the simplest and most suitable, because all the other cells of the body are derived from it, and because it is, in the strictest sense, the stem–father and stem–mother of all the countless generations of cells of which the multicellular organism is to be composed. That complicated molecular movement of the protoplasm which we call "life" is, naturally, something quite different in this stem–cell from what we find in the two parent–cells, from the coalescence of which it has issued. THE LIFE OF THE STEM–CELL OR CYTULA IS THE PRODUCT OR RESULTANT OF THE PATERNAL LIFE–MOVEMENT THAT IS CONVEYED IN THE SPERMATOZOON AND THE MATERNAL LIFE–MOVEMENT THAT IS CONTRIBUTED BY THE OVUM.

The admirable work done by recent observers has shown that the individual development, in man and the other animals, commences with the formation of a simple "stem–cell" of this character, and that this then passes, by repeated segmentation (or cleavage), into a cluster of cells, known as "the segmentation sphere" or "segmentation cells." The process is most clearly observed in the ova of the echinoderms (star–fishes, sea–urchins, etc.). The investigations of Oscar and Richard Hertwig were chiefly directed to these. The main results may be summed up as follows:—

The Evolution of Man, V.1.

Conception is preceded by certain preliminary changes, which are very necessary—in fact, usually indispensable—for its occurrence. They are comprised under the general heading of "Changes prior to impregnation." In these the original nucleus of the ovum, the germinal vesicle, is lost. Part of it is extruded, and part dissolved in the cell contents; only a very small part of it is left to form the basis of a fresh nucleus, the pronucleus femininus. It is the latter alone that combines in conception with the invading nucleus of the fertilising spermatozoon (the pronucleus masculinus).

The impregnation of the ovum commences with a decay of the germinal vesicle, or the original nucleus of the ovum (Figure 1.8). We have seen that this is in most unripe ova a large, transparent, round vesicle. This germinal vesicle contains a viscous fluid (the caryolymph). The firm nuclear frame (caryobasis) is formed of the enveloping membrane and a mesh−work of nuclear threads running across the interior, which is filled with the nuclear sap. In a knot of the network is contained the dark, stiff, opaque nuclear corpuscle or nucleolus. When the impregnation of the ovum sets in, the greater part of the germinal vesicle is dissolved in the cell; the nuclear membrane and mesh−work disappear; the nuclear sap is distributed in the protoplasm; a small portion of the nuclear base is extruded; another small portion is left, and is converted into the secondary nucleus, or the female pro−nucleus (Figure 1.24 e k).

The small portion of the nuclear base which is extruded from the impregnated ovum is known as the "directive bodies" or "polar cells"; there are many disputes as to their origin and significance, but we are as yet imperfectly acquainted with them. As a rule, they are two small round granules, of the same size and appearance as the remaining pro−nucleus. They are detached cell−buds; their separation from the large mother−cell takes place in the same way as in ordinary "indirect cell−division." Hence, the polar cells are probably to be conceived as "abortive ova," or "rudimentary ova," which proceed from a simple original ovum by cleavage in the same way that several sperm−cells arise from one "sperm−mother−cell," in reproduction from sperm. The male sperm−cells in the testicles must undergo similar changes in view of the coming impregnation as the ova in the female ovary. In this maturing of the sperm each of the original seed−cells divides by double segmentation into four daughter−cells, each furnished with a fourth of the original nuclear matter (the hereditary chromatin); and each of these four descendant cells becomes a spermatozoon, ready for impregnation. Thus is prevented the doubling of the chromatin in the coalescence of the two nuclei at conception. As the two polar cells are extruded and lost, and have no further part in the fertilisation of the ovum, we need not discuss them any further. But we must give more attention to the female pro—nucleus which alone remains after the extrusion of the polar cells and the dissolving of the germinal vesicle (Figure 1.23 e k). This tiny round corpuscle of chromatin now acts as a centre of attraction for the invading spermatozoon in the large ripe ovum, and coalesces with its "head," the male pro—nucleus. The product of this blending, which is the most important part of the act of impregnation, is the stem—nucleus, or the first segmentation nucleus (archicaryon)—that is to say, the nucleus of the new—born embryonic stem—cell or "first segmentation cell." This stem—cell is the starting point of the subsequent embryonic processes.

Hertwig has shown that the tiny transparent ova of the echinoderms are the most convenient for following the details of this important process of impregnation. We can, in this case, easily and successfully accomplish artificial impregnation, and follow the formation of the stem—cell step by step within the space of ten minutes. If we put ripe ova of the star—fish or sea—urchin in a watch glass with sea—water and add a drop of ripe sperm—fluid, we find each ovum impregnated within five minutes. Thousands of the fine, mobile ciliated cells, which we have described as "sperm—threads" (Figure 1.20), make their way to the ova, owing to a sort of chemical sensitive action which may be called "smell." But only one of these innumerable spermatozoa is chosen—namely, the one that first reaches the ovum by the serpentine motions of its tail, and touches the ovum with its head. At the spot where the point of its head touches the surface of the ovum the protoplasm of the latter is raised in the form of a small wart, the "impregnation rise" (Figure 1.25 A). The spermatozoon then bores its way into this with its head, the tail outside wriggling about all the time (Figure 1.25 B, C). Presently the tail also disappears within the ovum. At the same time the ovum secretes a thin external yelk—membrane (Figure 1.25 C), starting from the point of impregnation; and this prevents any more spermatozoa from entering.

Inside the impregnated ovum we now see a rapid series of most important changes. The pear—shaped head of the sperm—cell, or the "head of the spermatozoon," grows larger and rounder, and is converted into the male pro—nucleus (Figure 1.26 s k). This has an attractive influence on the fine granules or particles which are distributed in the protoplasm of the ovum; they arrange themselves in lines in the figure of a star. But the attraction or the "affinity" between the two nuclei is even stronger. They move towards each other inside the yelk with increasing speed, the male (Figure 1.27 s k) going more quickly than the female nucleus (e k). The tiny male nucleus takes with it the radiating mantle which spreads like a star about it. At last the two sexual nuclei touch (usually in the centre of the globular ovum), lie close together, are flattened at the points of contact, and coalesce into a common mass. The small central particle of nuclein which is formed from this combination of the nuclei is the stem–nucleus, or the first segmentation nucleus; the new–formed cell, the product of the impregnation, is our stem–cell, or "first segmentation sphere" (Figure 1.2).

(FIGURE 1.25. Impregnation of the ovum of a star–fish. (From Hertwig.) Only a small part of the surface of the ovum is shown. One of the numerous spermatozoa approaches the "impregnation rise" (A), touches it (B), and then penetrates into the protoplasm of the ovum (C).

FIGURES 1.26 AND 1.27. Impregnation of the ovum of the sea–urchin. (From Hertwig.) In Figure 1.26 the little sperm–nucleus (sk) moves towards the larger nucleus of the ovum (ek). In Figure 1.27 they nearly touch, and are surrounded by the radiating mantle of protoplasm.)

Hence the one essential point in the process of sexual reproduction or impregnation is the formation of a new cell, the stem–cell, by the combination of two originally different cells, the female ovum and the male spermatozoon. This process is of the highest importance, and merits our closest attention; all that happens in the later development of this first cell and in the life of the organism that comes of it is determined from the first by the chemical and morphological composition of the stem–cell, its nucleus and its body. We must, therefore, make a very careful study of the rise and structure of the stem–cell.

The first question that arises is as to the two different active elements, the nucleus and the protoplasm, in the actual coalescence. It is obvious that the nucleus plays the more important part in this. Hence Hertwig puts his theory of conception in the principle: "Conception consists in the copulation of two cell–nuclei, which come from a male and a female cell." And as the phenomenon of heredity is inseparably connected with the reproductive process, we may further conclude that these two copulating nuclei "convey the characteristics which are transmitted from parents to offspring." In this sense I had in 1866 (in the ninth chapter of the General Morphology) ascribed to the reproductive nucleus the function of generation and heredity, and to the nutritive protoplasm the duties of nutrition and adaptation. As, moreover, there is a complete coalescence of the mutually attracted nuclear substances in conception, and the new nucleus formed (the stem–nucleus) is the real starting–point for the development of the fresh organism, the further conclusion may be drawn that the male nucleus conveys to the child the qualities of the father, and the female nucleus the features of the mother. We must not forget, however, that the protoplasmic bodies of the copulating cells also fuse together in the act of impregnation; the cell–body of the invading spermatozoon (the trunk and tail of the male ciliated cell) is dissolved in the yelk of the female ovum. This coalescence is not so important as that of the nuclei, but it must not be overlooked; and, though this process is not so well known to us, we see clearly at least the formation of the star–like figure (the radial arrangement of the particles in the plasma) in it (Figures 1.26 to 1.27).

The older theories of impregnation generally went astray in regarding the large ovum as the sole base of the new organism, and only ascribed to the spermatozoon the work of stimulating and originating its development. The stimulus which it gave to the ovum was sometimes thought to be purely chemical, at other times rather physical (on the principle of transferred movement), or again a mystic and transcendental process. This error was partly due to the imperfect knowledge at that time of the facts of impregnation, and partly to the striking difference in the sizes of the two sexual cells. Most of the earlier observers thought that the spermatozoon did not penetrate into the ovum. And even when this had been demonstrated, the spermatozoon was believed to disappear in the ovum without leaving a trace. However, the splendid research made in the last three decades with the finer technical methods of our time has completely exposed the error of this. It has been shown that the tiny sperm–cell is NOT SUBORDINATED TO, BUT COORDINATED WITH, the large ovum. The nuclei of the two cells, as the vehicles of the hereditary features of the parents, are of equal physiological importance. In some cases we have succeeded in proving that the mass of the active nuclear substance which combines in the copulation of the two sexual nuclei is originally the same for both.

These morphological facts are in perfect harmony with the familiar physiological truth that the child inherits from both parents, and that on the average they are equally distributed. I say "on the average," because it is well known that a child may have a greater likeness to the father or to the mother; that goes without saying, as far as the primary sexual characters (the sexual glands) are concerned. But it is also possible that the determination of the latter—the weighty determination whether the child is to be a boy or a girl—depends on a slight qualitative or quantitative difference in the nuclein or the coloured nuclear matter which comes from both parents in the act of conception.

The striking differences of the respective sexual cells in size and shape, which occasioned the erroneous views of earlier scientists, are easily explained on the principle of division of labour. The inert, motionless ovum grows in size according to the quantity of provision it stores up in the form of nutritive yelk for the development of the germ. The active swimming sperm−cell is reduced in size in proportion to its need to seek the ovum and bore its way into its yelk. These differences are very conspicuous in the higher animals, but they are much less in the lower animals. In those protists (unicellular plants and animals) which have the first rudiments of sexual reproduction the two copulating cells are at first quite equal. In these cases the act of impregnation is nothing more than a sudden GROWTH, in which the originally simple cell doubles its volume, and is thus prepared for reproduction (cell−division). Afterwards slight differences are seen in the size of the copulating cells; though the smaller ones still have the same shape as the larger ones. It is only when the difference in size is very pronounced that a notable difference in shape is found: the sprightly sperm−cell changes more in shape and the ovum in size.

Quite in harmony with this new conception of the EQUIVALENCE OF THE TWO GONADS, or the equal physiological importance of the male and female sex−cells and their equal share in the process of heredity, is the important fact established by Hertwig (1875), that in normal impregnation only one single spermatozoon copulates with one ovum; the membrane which is raised on the surface of the yelk immediately after one sperm−cell has penetrated (Figure 1.25 C) prevents any others from entering. All the rivals of the fortunate penetrator are excluded, and die without. But if the ovum passes into a morbid state, if it is made stiff by a lowering of its temperature or stupefied with narcotics (chloroform, morphia, nicotine, etc.), two or more spermatozoa may penetrate into its yelk−body. We then witness polyspermism. The more Hertwig chloroformed the ovum, the more spermatozoa were able to bore their way into its unconscious body.

(FIGURE 1.28. Stem−cell of a rabbit, magnified 200 times. In the centre of the granular protoplasm of the fertilised ovum (d) is seen the little, bright stem−nucleus, z is the ovolemma, with a mucous membrane (h). s are dead spermatozoa.)

These remarkable facts of impregnation are also of the greatest interest in psychology, especially as regards the theory of the cell−soul, which I consider to be its chief foundation. The phenomena we have described can only be understood and explained by ascribing a certain lower degree of psychic activity to the sexual principles. They FEEL each other's proximity, and are drawn together by a SENSITIVE impulse (probably related to smell); they MOVE towards each other, and do not rest until they fuse together. Physiologists may say that it is only a question of a peculiar physico–chemical phenomenon, and not a psychic action; but the two cannot be separated. Even the psychic functions, in the strict sense of the word, are only complex physical processes, or "psycho–physical" phenomena, which are determined in all cases exclusively by the chemical composition of their material substratum.

The monistic view of the matter becomes clear enough when we remember the radical importance of impregnation as regards heredity. It is well known that not only the most delicate bodily structures, but also the subtlest traits of mind, are transmitted from the parents to the children. In this the chromatic matter of the male nucleus is just as important a vehicle as the large caryoplasmic substance of the female nucleus; the one transmits the mental features of the father, and the other those of the mother. The blending of the two parental nuclei determines the individual psychic character of the child.

But there is another important psychological question—the most important of all—that has been definitely answered by the recent discoveries in connection with conception. This is the question of the immortality of the soul. No fact throws more light on it and refutes it more convincingly than the elementary process of conception that we have described. For this copulation of the two sexual nuclei (Figures 1.26 and 1.27) indicates the precise moment at which the individual begins to exist. All the bodily and mental features of the new–born child are the sum–total of the hereditary qualities which it has received in reproduction from parents and ancestors. All that man acquires afterwards in life by the exercise of his organs, the influence of his environment, and education—in a word, by adaptation—cannot obliterate that general outline of his being which he inherited from his parents. But this hereditary disposition, the essence of every human soul, is not "eternal," but "temporal"; it comes into being only at the moment when the sperm–nucleus of the father and the nucleus of the maternal ovum meet and fuse together. It is clearly irrational to assume an "eternal life without end" for an individual phenomenon, the commencement of which we can indicate to a moment by direct visual observation.

The great importance of the process of impregnation in answering such questions is quite clear. It is true that conception has never been studied microscopically in all its details in the human case—notwithstanding its occurrence at every moment—for reasons that are obvious enough. However, the two cells which need consideration, the female ovum and the male spermatozoon, proceed in the case of man in just the same way as in all the other mammals; the human foetus or embryo which results from copulation has the same form as with the other animals. Hence, no scientist who is acquainted with the facts doubts that the processes of impregnation are just the same in man as in the other animals.

The stem–cell which is produced, and with which every man begins his career, cannot be distinguished in appearance from those of other mammals, such as the rabbit (Figure 1.28). In the case of man, also, this stem–cell differs materially from the original ovum, both in regard to form (morphologically), in regard to material composition (chemically), and in regard to vital properties (physiologically). It comes partly from the father and partly from the mother. Hence it is not surprising that the child who is developed from it inherits from both parents. The vital movements of each of these cells form a sum of mechanical processes which in the last analysis are due to movements of the smallest vital parts, or the molecules, of the living substance. If we agree to call this active substance plasson, and its molecules plastidules, we may say that the individual physiological character of each of these cells is due to its molecular plastidule–movement. HENCE, THE PLASTIDULE–MOVEMENT OF THE CYTULA IS THE RESULTANT OF THE COMBINED PLASTIDULE–MOVEMENTS OF THE FEMALE OVUM AND THE MALE SPERM–CELL.* (* The plasson of the stem–cell or cytula may, from the anatomical point of view, be regarded as homogeneous and structureless, like that of the monera. This is not inconsistent with our hypothetical ascription to the plastidules (or molecules of the plasson) of a complex molecular structure. The complexity of this is the greater in proportion to the complexity of the organism that is developed from it and the length of the chain of its ancestry, or to the multitude of antecedent processes of heredity and adaptation.)

CHAPTER 1.8. THE GASTRAEA THEORY.

There is a substantial agreement throughout the animal world in the first changes which follow the impregnation of the ovum and the formation of the stem–cell; they begin in all cases with the segmentation of the ovum and the formation of the germinal layers. The only exception is found in the protozoa, the very lowest and simplest forms of animal life; these remain unicellular throughout life. To this group belong the amoebae, gregarinae, rhizopods, infusoria, etc. As their whole organism consists of a single cell, they can never form germinal layers, or definite strata of cells. But all the other animals—all the tissue–forming animals, or metazoa, as we call them, in contradistinction to the protozoa—construct real germinal layers by the repeated cleavage of the impregnated ovum. This we find in the lower cnidaria and worms, as well as in the more highly–developed molluscs, echinoderms, articulates, and vertebrates.

In all these metazoa, or multicellular animals, the chief embryonic processes are substantially alike, although they often seem to a superficial observer to differ considerably. The stem–cell that proceeds from the impregnated ovum always passes by repeated cleavage into a number of simple cells. These cells are all direct descendants of the stem–cell, and are, for reasons we shall see presently, called segmentation–cells. The repeated cleavage of the stem–cell, which gives rise to these segmentation–spheres, has long been known as "segmentation." Sooner or later the segmentation–cells join together to form a round (at first, globular) embryonic sphere (blastula); they then form into two very different groups, and arrange themselves in two separate strata—the two primary germinal layers. These enclose a digestive cavity, the primitive gut, with an opening, the primitive mouth. We give the name of the gastrula to the important embryonic form that has these primitive organs, and the name of gastrulation to the formation of it. This ontogenetic process has a very great significance, and is the real starting–point of the construction of the multicellular animal body.

The fundamental embryonic processes of the cleavage of the ovum and the formation of the germinal layers have been very thoroughly studied in the last thirty years, and their real significance has been appreciated. They present a striking variety in the different groups, and it was no light task to prove their essential identity in the whole animal world. But since I formulated the gastraea theory in 1872, and afterwards (1875) reduced all the various forms of segmentation and gastrulation to one fundamental type, their identity may be said to have been established. We have thus mastered the law of unity which governs the first embryonic processes in all the animals.

Man is like all the other higher animals, especially the apes, in regard to these earliest and most important processes. As the human embryo does not essentially differ, even at a much later stage of development—when we already perceive the cerebral vesicles, the eyes, ears, gill–arches, etc.—from the similar forms of the other higher mammals, we may confidently assume that they agree in the earliest embryonic processes, segmentation and the formation of germinal layers. This has not yet, it is true, been established by observation. We have never yet had occasion to dissect a woman immediately after impregnation and examine the stem–cell or the segmentation–cells in her oviduct. However, as the earliest human embryos we have examined, and the later and more developed forms, agree with those of the rabbit, dog, and other higher mammals, no reasonable man will doubt but that the segmentation and formation of layers are the same in both cases.

But the special form of segmentation and layer formation which we find in the mammal is by no means the original, simple, palingenetic form. It has been much modified and cenogenetically altered by a very complex adaptation to embryonic conditions. We cannot, therefore, understand it altogether in itself. In order to do this, we have to make a COMPARATIVE study of segmentation and layer–formation in the animal world; and we have especially to seek the original, PALINGENETIC form from which the modified CENOGENETIC (see Chapter 1.1) form has gradually been developed.

This original unaltered form of segmentation and layer–formation is found to–day in only one case in the vertebrate–stem to which man belongs—the lowest and oldest member of the stem, the wonderful lancelet or amphioxus (cf. Chapters 2.16 and 2.17). But we find a precisely similar palingenetic form of embryonic development in the case of many of the invertebrate animals, as, for instance, the remarkable ascidia, the pond–snail (Limnaeus), and arrow–worm (Sagitta), and many of the echinoderms and cnidaria, such as the common star–fish and sea–urchin, many of the medusae and corals, and the simpler sponges (Olynthus). We may take as an illustration the palingenetic segmentation and germinal layer–formation in an eight–fold insular coral, which I discovered in the Red Sea, and described as Monoxenia Darwinii.

(FIGURE 1.29. Gastrulation of a coral (Monoxenia Darwinii). A, B, stem–cell (cytula) or impregnated ovum. In Figure A (immediately after impregnation) the nucleus is invisible. In Figure B (a little later) it is quite clear. C two segmentation–cells. D four segmentation–cells. E mulberry–formation (morula). F blastosphere (blastula). G blastula (transverse section). H depula, or hollowed blastula (transverse section). I gastrula (longitudinal section). K gastrula, or cup–sphere, external appearance.)

The impregnated ovum of this coral (Figure 1.29 A, B) first splits into two equal cells (C). First, the nucleus of the stem–cell and its central body divide into two halves. These recede from and repel each other, and act as centres of attraction on the surrounding protoplasm; in consequence of this, the protoplasm is constricted by a circular furrow, and, in turn, divides into two halves. Each of the two segmentation–cells thus produced splits in the same way into two equal cells. The four segmentation–cells (grand–daughters of the stem–cell) lie in one plane. Now, however, each of them subdivides into two equal halves, the cleavage of the nucleus again preceding that of the surrounding protoplasm. The eight cells which thus arise break into sixteen, these into thirty–two, and then (each being constantly halved) into sixty–four, 128, and so on.* (* The number of segmentation–cells thus produced increases geometrically in the original gastrulation, or the purest palingenetic form of cleavage. However, in different animals the number reaches a different height, so that the morula, and also the blastula, may consist sometimes of thirty–two, sometimes of sixty–four, and sometimes of 128, or more, cells.) The final result of this repeated cleavage is the formation of a globular cluster of similar segmentation–cells, which we call the mulberry–formation or morula. The cells are thickly pressed together like the parts of a mulberry or blackberry, and this gives a lumpy appearance to the surface of the sphere (Figure E).* (* The segmentation–cells which make up the morula after the close of the palingenetic cleavage seem usually to be quite similar, and to present no differences as to size, form, and composition. That, however, does not prevent them from differentiating into animal and vegetative cells, even during the cleavage.)

When the cleavage is thus ended, the mulberry–like mass changes into a hollow globular sphere. Watery fluid or jelly gathers inside the globule; the segmentation–cells are loosened, and all rise to the surface. There they are flattened by mutual pressure, and assume the shape of truncated pyramids, and arrange themselves side by side in one regular layer (Figures F, G). This layer of cells is called the germinal membrane (or blastoderm); the homogeneous cells which compose its simple structure are called blastodermic cells; and the whole hollow sphere, the walls of which are made of the preceding, is called the blastula or blastosphere.* (* The blastula of the lower animals must not be confused with the very different blastula of the mammal, which is properly called the gastrocystis or blastocystis. This cenogenetic gastrocystis and the palingenetic blastula are sometimes very wrongly comprised under the common name of blastula or vesicula blastodermica.)

In the case of our coral, and of many other lower forms of animal life, the young embryo begins at once to move independently and swim about in the water. A fine, long, thread–like process, a sort of whip or lash, grows out of each blastodermic cell, and this independently executes vibratory movements, slow at first, but quicker after a time (Figure F). In this way each blastodermic cell becomes a ciliated cell. The combined force of all these vibrating lashes causes the whole blastula to move about in a rotatory fashion. In many other animals, especially those in which the embryo develops within enclosed membranes, the ciliated cells are only formed at a later stage, or even not formed at all. The blastosphere may grow and expand by the blastodermic cells (at the surface of the sphere) dividing and increasing, and more fluid is secreted in the internal cavity. There are still to–day some organisms that remain throughout life at the structural stage of the blastula—hollow vesicles that swim about by a ciliary movement in the water, the wall of which is composed of a single layer of cells, such as the volvox, the magosphaera, synura, etc. We shall speak further of the great phylogenetic significance of this fact in Chapter 2.19.

A very important and remarkable process now follows—namely, the curving or invagination of the blastula (Figure H). The vesicle with a single layer of cells for wall is converted into a cup with a wall of two layers of cells (cf. Figures G, H, I). A certain spot at the surface of the sphere is flattened, and then bent inward. This depression sinks deeper and deeper, growing at the cost of the internal cavity. The latter decreases as the hollow deepens. At last the internal cavity disappears altogether, the inner side of the blastoderm (that which lines the depression) coming to lie close on the outer side. At the same time, the cells of the two sections assume different sizes and shapes; the inner cells are more round and the outer more oval (Figure I). In this way the embryo takes the form of a cup or jar–shaped body, with a wall made up of two layers of cells, the inner cavity of which opens to the outside at one end (the spot where the depression was originally formed). We call this very important and interesting embryonic form the "cup–embryo" or "cup–larva" (gastrula, Figure 1.29, I longitudinal section, K external view). I have in my Natural History of Creation given the name of depula to the remarkable intermediate form which appears at the passage of the blastula into the gastrula. In this intermediate stage there are two cavities in the embryo—the original cavity (blastocoel) which is disappearing, and the primitive gut–cavity (progaster) which is forming.

I regard the gastrula as the most important and significant embryonic form in the animal world. In all real animals (that is, excluding the unicellular protists) the segmentation of the ovum produces either a pure, primitive, palingenetic gastrula (Figure 1.29 I, K) or an equally instructive cenogenetic form, which has been developed in time from the first, and can be directly reduced to it. It is certainly a fact of the greatest interest and instructiveness that animals of the most different stems—vertebrates and tunicates, molluscs and articulates, echinoderms and annelids, cnidaria and sponges—proceed from one and the same embryonic form. In illustration I give a few pure gastrula forms from various groups of animals (Figures 1.30 to 1.35, explanation given below each).

(FIGURES 1.30 TO 1.35. In each figure d is the primitive–gut cavity, o primitive mouth, s segmentation–cavity, i entoderm (gut–layer), e ectoderm (skin layer).

FIGURE 1.30. (A) Gastrula of a very simple primitive–gut animal or gastraead (gastrophysema). (Haeckel.)

FIGURE 1.31. (B) Gastrula of a worm (Sagitta). (From Kowalevsky.)

FIGURE 1.32. (C) Gastrula of an echinoderm (star–fish, Uraster), not completely folded in (depula). (From Alexander Agassiz.)

FIGURE 1.33. (D) Gastrula of an arthropod (primitive crab, Nauplius) (as 32).

FIGURE 1.34. (E) Gastrula of a mollusc (pond–snail, Linnaeus). (From Karl Rabl.)

FIGURE 1.35. (F) Gastrula of a vertebrate (lancelet, Amphioxus). (From Kowalevsky.) (Front view.))

In view of this extraordinary significance of the gastrula, we must make a very careful study of its original structure. As a rule, the typical gastrula is very small, being invisible to the naked eye, or at the most only visible as a fine point under very favourable conditions, and measuring generally 1/500 to 1/250 of an inch (less frequently 1/50 inch, or even more) in diameter. In shape it is usually like a roundish drinking–cup. Sometimes it is rather oval, at other times more ellipsoid or spindle–shaped; in some cases it is half round, or even almost round, and in others lengthened out, or almost cylindrical.

I give the name of primitive gut (progaster) and primitive mouth (prostoma) to the internal cavity of the gastrula–body and its opening; because this cavity is the first rudiment of the digestive cavity of the organism, and the opening originally served to take food into it. Naturally, the primitive gut and mouth change very considerably afterwards in the various classes of animals. In most of the cnidaria and many of the annelids (worm–like animals) they remain unchanged throughout life. But in most of the higher animals, and so in the vertebrates, only the larger central part of the later alimentary canal develops from the primitive gut; the later mouth is a fresh development, the primitive mouth disappearing or changing into the anus. We must therefore distinguish carefully between the primitive gut and mouth of the gastrula and the later alimentary canal and mouth of the fully developed vertebrate.* (* My distinction (1872) between the primitive gut and mouth and the later permanent stomach (metagaster) and mouth (metastoma) has been much criticised; but it is as much justified as the distinction between the primitive kidneys and the permanent kidneys. Professor E. Ray–Lankester suggested three years afterwards (1875) the name archenteron for the primitive gut, and blastoporus for the primitive mouth.)

(FIGURE 1.36. Gastrula of a lower sponge (olynthus). A external view, B longitudinal section through the axis, g primitive–gut cavity, a primitive mouth–aperture, i inner cell–layer (entoderm, endoblast, gut–layer), e external cell–layer (outer germinal layer, ectoderm, ectoblast, or skin–layer).

The two layers of cells which line the gut–cavity and compose its wall are of extreme importance. These two layers, which are the sole builders of the whole organism, are no other than the two primary germinal layers, or the primitive germ–layers. I have spoken in the introductory section (Chapter 1.3.) of their radical importance. The outer stratum is the skin–layer, or ectoderm (Figures 1.30 to 1.35 e); the inner stratum is the gut–layer, or entoderm (i). The former is often also called the ectoblast, or epiblast, and the latter the endoblast, or hypoblast. FROM THESE TWO PRIMARY GERMINAL LAYERS ALONE IS DEVELOPED THE ENTIRE ORGANISM OF ALL THE METAZOA OR MULTICELLULAR ANIMALS. The skin–layer forms the external skin, the gut–layer forms the internal skin or lining of the body. Between these two germinal layers are afterwards developed the middle germinal layer (mesoderma) and the body–cavity (coeloma) filled with blood or lymph.

The two primary germinal layers were first distinguished by Pander in 1817 in the incubated chick. Twenty years later (1849) Huxley pointed out that in many of the lower zoophytes, especially the medusae, the whole body consists throughout life of these two primary germinal layers. Soon afterwards (1853) Allman introduced the names which have come into general use; he called the outer layer the ectoderm ("outer–skin"), and the inner the entoderm ("inner–skin"). But in 1867 it was shown, particularly by Kowalevsky, from comparative observation, that even in invertebrates, also, of the most different classes—annelids, molluscs, echinoderms, and articulates—the body is developed out of the same two primary layers. Finally, I discovered them (1872) in the lowest tissue–forming animals, the sponges, and proved in my gastraea theory that these two layers must be regarded as identical throughout the animal world, from the sponges and corals to the insects and vertebrates, including man. This fundamental "homology [identity] of the primary germinal layers and the primitive gut" has been confirmed during the last thirty years by the careful research of many able observers, and is now pretty generally admitted for the whole of the metazoa.

As a rule, the cells which compose the two primary germinal layers show appreciable differences even in the gastrula stage. Generally (if not always) the cells of the skin–layer or ectoderm (Figures 1.36 c and 1.37 e) are the smaller, more numerous, and clearer; while the cells of the gut–layer, or entoderm (i), are larger, less numerous, and darker. The protoplasm of the ectodermic (outer) cells is clearer and firmer than the thicker and softer cell–matter of the entodermic (inner) cells; the latter are, as a rule, much richer in yelk–granules (albumen and fatty particles) than the former. Also the cells of the gut–layer have, as a rule, a stronger affinity for colouring matter, and take on a tinge in a solution of carmine, aniline, etc., more quickly and appreciably than the cells of the skin–layer. The nuclei of the entoderm–cells are usually roundish, while those of the ectoderm–cells are oval.

When the doubling–process is complete, very striking histological differences between the cells of the two layers are found (Figure 1.37). The tiny, light ectoderm–cells (e) are sharply distinguished from the larger and darker entoderm–cells (i). Frequently this differentiation of the cell–forms sets in at a very early stage, during the segmentation–process, and is already very appreciable in the blastula.

We have, up to the present, only considered that form of segmentation and gastrulation which, for many and weighty reasons, we may regard as the original, primordial, or palingenetic form. We might call it "equal" or homogeneous segmentation, because the divided cells retain a resemblance to each other at first (and often until the formation of the blastoderm). We give the name of the "bell–gastrula," or archigastrula, to the gastrula that succeeds it. In just the same form as in the coral we considered (Monoxenia, Figure 1.29), we find it in the lowest zoophyta (the gastrophysema, Figure 1.30), and the simplest sponges (olynthus, Figure 1.36); also in many of the medusae and hydrapolyps, lower types of worms of various classes (brachiopod, arrow–worm, Figure 1.31), tunicates (ascidia), many of the echinoderms (Figure 1.32), lower articulates (Figure 1.33), and molluscs (Figure 1.34), and, finally, in a slightly modified form, in the lowest vertebrate (the amphioxus, Figure 1.35).

(FIGURE 1.37. Cells from the two primary germinal layers of the mammal (from both layers of the blastoderm). i larger and darker cells of the inner stratum, the vegetal layer or entoderm. e smaller and clearer cells from the outer stratum, the animal layer or ectoderm.

FIGURE 1.38. Gastrulation of the amphioxus, from Hatschek (vertical section through the axis of the ovum). A, B, C three stages in the formation of the blastula; D, E curving of the blastula; F complete gastrula. h segmentation–cavity. g primitive gut–cavity.))

The gastrulation of the amphioxus is especially interesting because this lowest and oldest of all the vertebrates is of the highest significance in connection with the evolution of the vertebrate stem, and therefore with that of man (compare Chapters 2.16 and 2.17). Just as the comparative anatomist traces the most elaborate features in the structures of the various classes of vertebrates to divergent development from this simple primitive vertebrate, so comparative embryology traces the various secondary forms of vertebrate gastrulation to the simple, primary formation of the germinal layers in the amphioxus. Although this formation, as distinguished from the cenogenetic modifications of the vertebrate, may on the whole be regarded as palingenetic, it is nevertheless different in some features from the quite primitive gastrulation such as we have, for instance, in the Monoxenia (Figure 1.29) and the Sagitta. Hatschek rightly observes that the segmentation of the ovum in the amphioxus is not strictly equal, but almost equal, and approaches the unequal. The difference in size between the two groups of cells continues to be very noticeable in the further course of the segmentation; the smaller animal cells of the upper hemisphere divide more quickly than the larger vegetal cells of the lower (Figure 1.38 A, B). Hence the blastoderm, which forms the single–layer wall of the globular blastula at the end of the cleavage–process, does not consist of homogeneous cells of equal size, as in the Sagitta and the Monoxenia; the cells of the upper half of the blastoderm (the mother–cells of the ectoderm) are more numerous and smaller, and the cells of the lower half (the mother–cells of the entoderm) less numerous and larger. Moreover, the segmentation–cavity of the blastula (Figure 1.38 C, h) is not quite globular, but forms a flattened spheroid with unequal poles of its vertical axis. While the blastula is being folded into a cup at the vegetal pole of its axis, the difference in the size of the blastodermic cells increases (Figure 1.38 D, E); it is most conspicuous when the invagination is complete and the segmentation–cavity has disappeared (Figure 1.38 F). The larger vegetal cells of the entoderm are richer in granules, and so darker than the smaller and lighter animal cells of the ectoderm.

But the unequal gastrulation of the amphioxus diverges from the typical equal cleavage of the Sagitta, the Monoxenia (Figure 1.29), and the Olynthus (Figure 1.36), in another important particular. The pure archigastrula of the latter forms is uni–axial, and it is round in its whole length in transverse section. The vegetal pole of the vertical axis is just in the centre of the primitive mouth. This is not the case in the gastrula of the amphioxus. During the folding of the blastula the ideal axis is already bent on one side, the growth of the blastoderm (or the increase of its cells) being brisker on one side than on the other; the side that grows more quickly, and so is more curved (Figure 1.39 v), will be the anterior or belly–side, the opposite, flatter side will form the back (d). The primitive mouth, which at first, in the typical archigastrula, lay at the vegetal pole of the main axis, is forced away to the dorsal side; and whereas its two lips lay at first in a plane at right angles to the chief axis, they are now so far thrust aside that their plane cuts the axis at a sharp angle. The dorsal lip is therefore the upper and more forward, the ventral lip the lower and hinder. In the latter, at the ventral passage of the entoderm into the ectoderm, there lie side by side a pair of very large cells, one to the right and one to the left (Figure 1.39 p): these are the important polar cells of the primitive mouth, or "the primitive cells of the mesoderm." In consequence of these considerable variations arising in the course of the gastrulation, the primitive uni–axial form of the archigastrula in the amphioxus has already become tri–axial, and thus the two–sidedness, or bilateral symmetry, of the vertebrate body has already been determined. This has been transmitted from the amphioxus to all the other modified gastrula–forms of the vertebrate stem.

Apart from this bilateral structure, the gastrula of the amphioxus resembles the typical archigastrula of the lower animals (Figures 1.30 to 1.36) in developing the two primary germinal layers from a single layer of cells. This is clearly the oldest and original form of the metazoic embryo. Although the animals I have mentioned belong to the most diverse classes, they nevertheless agree with each other, and many more animal forms, in having retained to the present day, by a conservative heredity, this palingenetic form of gastrulation which they have from their earliest common ancestors. But this is not the case with the great majority of the animals. With these the original embryonic process has been gradually more or less altered in the course of millions of years by adaptation to new conditions of development. Both the segmentation of the ovum and the subsequent gastrulation have in this way been considerably changed. In fact, these variations have become so great in the course of time that the segmentation was not rightly understood in most animals, and the gastrula was unrecognised. It was not until I had made an extensive comparative study, lasting a considerable time (in the years 1866 to 1875), in animals of the most diverse classes, that I succeeded in showing the same common typical process in these apparently very different forms of gastrulation, and tracing them all to one original form. I regard all those that diverge from the primary palingenetic gastrulation as secondary, modified, and cenogenetic. The more or less divergent form of gastrula that is produced may be called a secondary, modified gastrula, or a metagastrula. The reader will find a scheme of these different kinds of segmentation and gastrulation at the close of this chapter.

By far the most important process that determines the various cenogenetic forms of gastrulation is the change in the nutrition of the ovum and the accumulation in it of nutritive yelk. By this we understand various chemical substances (chiefly granules of albumin and fat–particles) which serve exclusively as reserve–matter or food for the embryo. As the metazoic embryo in its earlier stages of development is not yet able to obtain its food and so build up the frame, the necessary material has to be stored up in the ovum. Hence we distinguish in the ova two chief elements—the active formative yelk (protoplasm) and the passive food–yelk (deutoplasm, wrongly spoken of as "the yelk"). In the little palingenetic ova, the segmentation of which we have already considered, the yelk–granules are so small and so regularly distributed in the protoplasm of the ovum that the even and repeated cleavage is not affected by them. But in the great majority of the animal ova the food–yelk is more or less considerable, and is stored in a certain part of the ovum, so that even in the unfertilised ovum the "granary" can clearly be distinguished from the formative plasm. As a rule, the formative–yelk (with the germinal vesicle) then usually gathers at one pole and the food–yelk at the other. The first is the ANIMAL, and the second the VEGETAL, pole of the vertical axis of the ovum.

(FIGURE 1.39. Gastrula of the amphioxus, seen from left side (diagrammatic median section). (From Hatschek.) g primitive gut, u primitive mouth, p peristomal pole–cells, i entoderm, e ectoderm, d dorsal side, v ventral side.)

In these "telolecithal" ova, or ova with the yelk at one end (for instance, in the cyclostoma and amphibia), the gastrulation then usually takes place in such a way that in the cleavage of the impregnated ovum the animal (usually the upper) half splits up more quickly than the vegetal (lower). The contractions of the active protoplasm, which effect this continual cleavage of the cells, meet a greater resistance in the lower vegetal half from the passive deutoplasm than in the upper animal half. Hence we find in the latter more but smaller, and in the former fewer but larger, cells. The animal cells produce the external, and the vegetal cells the internal, germinal layer.

Although this unequal segmentation of the cyclostoma, ganoids, and amphibia seems at first sight to differ from the original equal segmentation (for instance, in the monoxenia, Figure 1.29), they both have this in common, that the cleavage process throughout affects the WHOLE cell; hence Remak called it TOTAL segmentation, and the ova in question holoblastic, or "whole–cleaving." It is otherwise with the second chief group of ova, which he distinguished from these as meroblastic, or "partially–cleaving ": to this class belong the familiar large eggs of birds and reptiles, and of most fishes. The inert mass of the passive food–yelk is so large in these cases that the protoplasmic contractions of the active yelk cannot effect any further cleavage. In consequence, there is only a partial segmentation. While the protoplasm in the animal section of the ovum continues briskly to divide, multiplying the nuclei, the deutoplasm in the vegetal section remains more or less undivided; it is merely consumed as food by the forming cells. The larger the accumulation of food, the more restricted is the process of segmentation. It may, however, continue for some time (even after the gastrulation is more or less complete) in the sense that the vegetal cell–nuclei distributed in the deutoplasm slowly increase by cleavage; as each of them is surrounded by a small quantity of protoplasm, it may afterwards appropriate a portion of the food–yelk, and thus form a real "yelk–cell" (merocyte). When this vegetal cell–formation continues for a long time, after the two primary germinal layers have been formed, it takes the name of the "after–segmentation."

The meroblastic ova are only found in the larger and more highly developed animals, and only in those whose embryo needs a longer time and richer nourishment within the foetal membranes. According as the yelk–food accumulates at the centre or at the side of the ovum, we distinguish two groups of dividing ova, periblastic and discoblastic. In the periblastic the food–yelk is in the centre, enclosed inside the ovum (hence they are also called "centrolecithal" ova): the formative yelk surrounds the food–yelk, and so suffers itself a superficial cleavage. This is found among the articulates (crabs, spiders, insects, etc.). In the discoblastic ova the food–yelk gathers at one side, at the vegetal or lower pole of the vertical axis, while the nucleus of the ovum and the great bulk of the formative yelk lie at the upper or animal pole (hence these ova are also called "telolecithal"). In these cases the cleavage of the ovum begins at the upper pole, and leads to the formation of a dorsal discoid embryo. This is the case with all meroblastic vertebrates, most fishes, the reptiles and birds, and the oviparous mammals (the monotremes).

The gastrulation of the discoblastic ova, which chiefly concerns us, offers serious difficulties to microscopic investigation and philosophic consideration. These, however, have been mastered by the comparative embryological research which has been conducted by a number of distinguished observers during the last few decades—especially the brothers Hertwig, Rabl, Kupffer, Selenka, Ruckert, Goette, Rauber, etc. These thorough and careful studies, aided by the most perfect modern improvements in technical method (in tinting and dissection), have given a very welcome support to the views which I put forward in my work, On the Gastrula and the Segmentation of the Animal Ovum [not translated], in 1875. As it is very important to understand these views and their phylogenetic foundation clearly, not only as regards evolution in general, but particularly in connection with the genesis of man, I will give here a brief statement of them as far as they concern the vertebrate–stem:—

1. All the vertebrates, including man, are phylogenetically (or genealogically) related—that is, are members of one single natural stem.

2. Consequently, the embryonic features in their individual development must also have a genetic connection.

3. As the gastrulation of the amphioxus shows the original palingenetic form in its simplest features, that of the other vertebrates must have been derived from it.

4. The cenogenetic modifications of the latter are more appreciable the more food–yelk is stored up in the ovum.

5. Although the mass of the food–yelk may be very large in the ova of the discoblastic vertebrates, nevertheless in every case a blastula is developed from the morula, as in the holoblastic ova.

6. Also, in every case, the gastrula develops from the blastula by curving or invagination.

7. The cavity which is produced in the foetus by this curving is, in each case, the primitive gut (progaster), and its opening the primitive mouth (prostoma).

8. The food–yelk, whether large or small, is always stored in the ventral wall of the primitive gut; the cells (called "merocytes") which may be formed in it subsequently (by "after–segmentation") also belong to the inner germinal layer, like the cells which immediately enclose the primitive gut–cavity.

9. The primitive mouth, which at first lies below at the lower pole of the vertical axis, is forced, by the growth of the yelk, backwards and then upwards, towards the dorsal side of the embryo; the vertical axis of the primitive gut is thus gradually converted into horizontal.

10. The primitive mouth is closed sooner or later in all the vertebrates, and does not evolve into the permanent mouth–aperture; it rather corresponds to the "properistoma," or region of the anus. From this important point the formation of the middle germinal layer proceeds, between the two primary layers.

The wide comparative studies of the scientists I have named have further shown that in the case of the discoblastic higher vertebrates (the three classes of amniotes) the primitive mouth of the embryonic disc, which was long looked for in vain, is found always, and is nothing else than the familiar "primitive groove." Of this we shall see more as we proceed. Meantime we realise that gastrulation may be reduced to one and the same process in all the vertebrates. Moreover, the various forms it takes in the invertebrates can always be reduced to one of the four types of segmentation described above. In relation to the distinction between total and partial segmentation, the grouping of the various forms is as follows:—

1. Palingenetic (primitive segmentation)

1.1. Equal segmentation (bell–gastrula).

1.1.A. Total segmentation (without independent food–yelk).

2. Cenogenetic segmentation (modified by adaptation).

2.2. Unequal segmentation (hooded gastrula).

2.2.A. Total segmentation (without independent food–yelk).

2.3. Discoid segmentation (discoid gastrula).

2.3.B. Partial segmentation (with independent food–yelk).

2.4. Superficial segmentation (spherical gastrula).

2.4.B. Partial segmentation (with independent food–yelk).

The lowest metazoa we know—namely, the lower zoophyta (sponges, simple polyps, etc.)—remain throughout life at a stage of development which differs little from the gastrula; their whole body consists of two layers of cells. This is a fact of extreme importance. We see that man, and also other vertebrates, pass quickly through a stage of development in which they consist of two layers, just as these lower zoophyta do throughout life. If we apply our biogenetic law to the matter, we at once reach this important conclusion. "Man and all the other animals which pass through the two–layer stage, or gastrula–form, in the course of their embryonic development, must descend from a primitive simple stem–form, the whole body of which consisted throughout life (as is the case with the lower zoophyta to–day) merely of two cell–strata or germinal layers." We will call this primitive stem–form, with which we shall deal more fully later on, the gastraea—that is to say, "primitive–gut animal."

According to this gastraea–theory there was originally in all the multicellular animals ONE ORGAN with the same structure and function. This was the primitive gut; and the two primary germinal layers which form its wall must also be regarded as identical in all. This important homology or identity of the primary germinal layers is proved, on the one hand, from the fact that the gastrula was originally formed in the same way in all

116

cases—namely, by the curving of the blastula; and, on the other hand, by the fact that in every case the same fundamental organs arise from the germinal layers. The outer or animal layer, or ectoderm, always forms the chief organs of animal life—the skin, nervous system, sense–organs, etc.; the inner or vegetal layer, or entoderm, gives rise to the chief organs of vegetative life—the organs of nourishment, digestion, blood–formation, etc.

In the lower zoophyta, whose body remains at the two–layer stage throughout life, the gastraeads, the simplest sponges (Olynthus), and polyps (Hydra), these two groups of functions, animal and vegetative, are strictly divided between the two simple primary layers. Throughout life the outer or animal layer acts simply as a covering for the body, and accomplishes its movement and sensation. The inner or vegetative layer of cells acts throughout life as a gut–lining, or nutritive layer of enteric cells, and often also yields the reproductive cells.

The best known of these "gastraeads," or "gastrula–like animals," is the common fresh–water polyp (Hydra). This simplest of all the cnidaria has, it is true, a crown of tentacles round its mouth. Also its outer germinal layer has certain special modifications. But these are secondary additions, and the inner germinal layer is a simple stratum of cells. On the whole, the hydra has preserved to our day by heredity the simple structure of our primitive ancestor, the gastraea (cf. Chapter 2.19.)

In all other animals, particularly the vertebrates, the gastrula is merely a brief transitional stage. Here the two–layer stage of the embryonic development is quickly succeeded by a three–layer, and then a four–layer, stage. With the appearance of the four superimposed germinal layers we reach again a firm and steady standing–ground, from which we may follow the further, and much more difficult and complicated, course of embryonic development.

SUMMARY OF THE CHIEF DIFFERENCES IN THE OVUM–SEGMENTATION AND GASTRULATION OF ANIMALS.

The animal stems are indicated by the letters a–g: a Zoophyta. b Annelida. c Mollusca.
d Echinoderma. e Articulata. f Tunicata. g Vertebrata.

1. Total Segmentation. Holoblastic ova. Gastrula without separate food–yelk. Hologastrula.

1.1. Primitive Segmentation. Archiblastic ova. Bell–gastrula (archigastrula.)
a. Many lower zoophyta (sponges, hydrapolyps, medusae, simpler corals).
b. Many lower annelids (sagitta, phoronis, many nematoda, etc., terebratula, argiope, pisidium).
c. Some lower molluscs.
d. Many echinoderms.
e. A few lower articulata (some brachiopods, copepods: Tardigrades, pteromalina).
f. Many tunicata.
g. The acrania (amphioxus).

1.2. Unequal Segmentation. Amphiblastic ova. Hooded–gastrula (amphigastrula).
a. Many zoophyta (sponges, medusae, corals, siphonophorae, ctenophora).
b. Most worms.
c. Most molluscs.
d. Many echinoderms (viviparous species and some others).
e. Some of the lower articulata (both crustacea and tracheata).
f. Many tunicata.
g. Cyclostoma, the oldest fishes, amphibia, mammals (not including man).

2. Partial Segmentation. Meroblastic ova. Gastrula with separate food–yelk. Merogastrula.

2.3. Discoid Segmentation. Discoblastic ova. Discoid gastrula.
c. Cephalopods or cuttlefish.
e. Many articulata, wood–lice, scorpions, etc.
g. Primitive fishes, bony fishes, reptiles, birds, monotremes.

2.4. Superficial Segmentation. Periblastic ova. Spherical–gastrula.
e. The great majority of the articulata (crustaceans, myriapods,
arachnids, insects).

CHAPTER 1.9. THE GASTRULATION OF THE VERTEBRATE.*

(* Cf. Balfour's Manual of Comparative Embryology volume 2; Theodore Morgan's The Development of the Frog's Egg.)

The remarkable processes of gastrulation, ovum–segmentation, and formation of germinal layers present a most conspicuous variety. There is to–day only the lowest of the vertebrates, the amphioxus, that exhibits the original form of those processes, or the palingenetic gastrulation which we have considered in the preceding chapter, and which culminates in the formation of the archigastrula (Figure 1.38). In all other extant vertebrates these fundamental processes have been more or less modified by adaptation to the conditions of embryonic development (especially by changes in the food–yelk); they exhibit various cenogenetic types of the formation of germinal layers. However, the different classes vary considerably from each other. In order to grasp the unity that underlies the manifold differences in these phenomena and their historical connection, it is necessary to bear in mind always the unity of the vertebrate–stem. This "phylogenetic unity," which I developed in my General Morphology in 1866, is now generally admitted. All impartial zoologists agree to–day that all the vertebrates, from the amphioxus and the fishes to the ape and man, descend from a common ancestor, "the primitive vertebrate." Hence the embryonic processes, by which each individual vertebrate is developed, must also be capable of being reduced to one common type of embryonic development; and this primitive type is most certainly exhibited to–day by the amphioxus.

It must, therefore, be our next task to make a comparative study of the various forms of vertebrate gastrulation, and trace them backwards to that of the lancelet. Broadly speaking, they fall first into two groups: the older cyclostoma, the earliest fishes, most of the amphibia, and the viviparous mammals, have holoblastic ova—that is to say, ova with total, unequal segmentation; while the younger cyclostoma, most of the fishes, the cephalopods, reptiles, birds, and monotremes, have meroblastic ova, or ova with partial discoid segmentation. A closer study of them shows, however, that these two groups do not present a natural unity, and that the historical relations between their several divisions are very complicated. In order to understand them properly, we must first consider the various modifications of gastrulation in these classes. We may begin with that of the amphibia.

The most suitable and most available objects of study in this class are the eggs of our indigenous amphibia, the tailless frogs and toads, and the tailed salamander. In spring they are to be found in clusters in every pond, and careful examination of the ova with a lens is sufficient to show at least the external features of the segmentation. In order to understand the whole process rightly and follow the formation of the germinal layers and the gastrula, the ova of the frog and salamander must be carefully hardened; then the thinnest possible sections must be made of the hardened ova with the microtome, and the tinted sections must be very closely compared under a powerful microscope.

The ova of the frog or toad are globular in shape, about the twelfth of an inch in diameter, and are clustered in jelly–like masses, which are lumped together in the case of the frog, but form long strings in the case of the toad. When we examine the opaque, grey, brown, or blackish ova closely, we find that the upper half is darker than the lower. The middle of the upper half is in many species black, while the middle of the lower half is white.* (* The colouring of the eggs of the amphibia is caused by the accumulation of dark–colouring matter at the animal pole of the ovum. In consequence of this, the animal cells of the ectoderm are darker than the vegetal cells of the entoderm. We find the reverse of this in the case of most animals, the protoplasm of the entoderm cells being usually darker and coarser–grained.) In this way we get a definite axis of the ovum with two poles. To give a clear idea of the segmentation of this ovum, it is best to compare it with a globe, on the surface of which are marked the various parallels of longitude and latitude. The superficial dividing lines between the different cells, which come from the repeated segmentation of the ovum, look like deep furrows on the surface, and hence the whole process has been given the name of furcation. In reality, however, this "furcation," which was formerly regarded as a very mysterious process, is nothing but the familiar, repeated cell–segmentation. Hence also the segmentation–cells which result from it are real cells.

(FIGURE 1.40. The cleavage of the frog's ovum (magnified ten times). A stem–cell. B the first two segmentation–cells. C four cells. D eight cells (4 animal and 4 vegetative). E twelve cells (8 animal and 4 vegetative). F sixteen cells (8 animal and 8 vegetative). G twenty–four cells (16 animal and 8 vegetative). H thirty–two cells. I forty–eight cells. K sixty–four cells. L ninety–six cells. M 160 cells (128 animal and 32 vegetative).

(FIGURES 1.41 TO 1.44. Four vertical sections of the fertilised ovum of the toad, in four successive stages of development. The letters have the same meaning throughout: F segmentation–cavity. D covering of same (D dorsal half of the embryo, P ventral half). P yelk–stopper (white round field at the lower pole). Z yelk–cells of the entoderm (Remak's "glandular embryo"). N primitive gut cavity (progaster or Rusconian alimentary cavity). The primitive mouth (prostoma) is closed by the yelk–stopper, P. s partition between the primitive gut cavity (N) and the segmentation cavity (F). k k apostrophe, section of the large circular lip–border of the primitive mouth (the Rusconian anus). The line of dots between k and k apostrophe indicates the earlier connection of the yelk–stopper (P) with the central mass of the yelk–cells (Z). In Figure 1.44 the ovum has turned 90 degrees, so that the back of the embryo is uppermost and the ventral side down. (From Stricker.)).

The unequal segmentation which we observe in the ovum of the amphibia has the special feature of beginning at the upper and darker pole (the north pole of the terrestrial globe in our illustration), and slowly advancing towards the lower and brighter pole (the south pole). Also the upper and darker hemisphere remains in this position throughout the course of the segmentation, and its cells multiply much more briskly. Hence the cells of the lower hemisphere are found to be larger and less numerous. The cleavage of the stem–cell (Figure 1.40 A) begins with the formation of a complete furrow, which starts from the north pole and reaches to the south (B). An hour later a second furrow arises in the same way, and this cuts the first at a right angle (Figure 1.40 C). The ovum is thus divided into four equal parts. Each of these four "segmentation cells" has an upper and darker and a lower, brighter half. A few hours later a third furrow appears, vertically to the first two (Figure 1.40 D). The globular germ now consists of eight cells, four smaller ones above (northern) and four larger ones below (southern). Next, each of the four upper ones divides into two halves by a cleavage beginning from the north pole, so that we now have eight above and four below (Figure 1.40 E). Later, the four new longitudinal divisions extend gradually to the lower cells, and the number rises from twelve to sixteen (F). Then a second circular furrow appears, parallel to the first, and nearer to the north pole, so that we may compare it to the north polar circle. In this way we get twenty–four segmentation–cells—sixteen upper, smaller, and darker ones, and eight smaller and

121

brighter ones below (G). Soon, however, the latter also sub–divide into sixteen, a third or "meridian of latitude" appearing, this time in the southern hemisphere: this makes thirty–two cells altogether (H). Then eight new longitudinal lines are formed at the north pole, and these proceed to divide, first the darker cells above and afterwards the lighter southern cells, and finally reach the south pole. In this way we get in succession forty, forty–eight, fifty–six, and at last sixty–four cells (I, K). In the meantime, the two hemispheres differ more and more from each other. Whereas the sluggish lower hemisphere long remains at thirty–two cells, the lively northern hemisphere briskly sub–divides twice, producing first sixty–four and then 128 cells (L, M). Thus we reach a stage in which we count on the surface of the ovum 128 small cells in the upper half and thirty–two large ones in the lower half, or 160 altogether. The dissimilarity of the two halves increases: while the northern breaks up into a great number of small cells, the southern consists of a much smaller number of larger cells. Finally, the dark cells of the upper half grow almost over the surface of the ovum, leaving only a small circular spot at the south pole, where the large and clear cells of the lower half are visible. This white region at the south pole corresponds, as we shall see afterwards, to the primitive mouth of the gastrula. The whole mass of the inner and larger and clearer cells (including the white polar region) belongs to the entoderm or ventral layer. The outer envelope of dark smaller cells forms the ectoderm or skin–layer.

In the meantime, a large cavity, full of fluid, has been formed within the globular body—the segmentation–cavity or embryonic cavity (blastocoel, Figures 1.41 to 1.44 F). It extends considerably as the cleavage proceeds, and afterwards assumes an almost semi–circular form (Figure 1.41 F). The frog–embryo now represents a modified embryonic vesicle or blastula, with hollow animal half and solid vegetal half.

Now a second, narrower but longer, cavity arises by a process of folding at the lower pole, and by the falling away from each other of the white entoderm–cells (Figures 1.41 to 1.44 N). This is the primitive gut–cavity or the gastric cavity of the gastrula, progaster or archenteron. It was first observed in the ovum of the amphibia by Rusconi, and so called the Rusconian cavity. The reason of its peculiar narrowness here is that it is, for the most part, full of yelk–cells of the entoderm. These also stop up the whole of the wide opening of the primitive mouth, and form what is known as the "yelk–stopper," which is seen freely at the white round spot at the south pole (P). Around it the ectoderm is much thicker, and forms the border of the primitive mouth, the most important part of the embryo (Figure 1.44 k, k apostrophe). Soon the primitive gut–cavity stretches further and further at the expense of the segmentation–cavity (F), until at last the latter disappears altogether. The two cavities are only separated by a thin partition (Figure 1.43 s). With the formation of the primitive gut our frog–embryo has reached the gastrula stage, though it is clear that this cenogenetic amphibian gastrula is very different from the real palingenetic gastrula we have considered (Figures 1.30 to 1.36).

In the growth of this hooded gastrula we cannot sharply mark off the various stages which we distinguish successively in the bell–gastrula as morula and gastrula. Nevertheless, it is not difficult to reduce the whole cenogenetic or disturbed development of this amphigastrula to the true palingenetic formation of the archigastrula of the amphioxus.

(FIGURE 1.45. Blastula of the water–salamander (Triton). fh segmentation–cavity, dz yelk–cells, rz border–zone. (From Hertwig.)

FIGURE 1.46. Embryonic vesicle of triton (blastula), outer view, with the transverse fold of the primitive mouth (u). (From Hertwig.)

FIGURE 1.47. Sagittal section of a hooded–embryo (depula) of triton (blastula at the commencement of gastrulation). ak outer germinal layer, ik inner germinal layer, fh segmentation–cavity, ud primitive gut, u primitive mouth, dl and vl dorsal and ventral lips of the mouth, dz yelk–cells. (From Hertwig.))

This reduction becomes easier if, after considering the gastrulation of the tailless amphibia (frogs and toads), we glance for a moment at that of the tailed amphibia, the salamanders. In some of the latter, that have only recently been carefully studied, and that are phylogenetically older, the process is much simpler and clearer than is the case with the former and longer known. Our common water–salamander (Triton taeniatus) is a particularly good subject for observation. Its nutritive yelk is much smaller and its formative yelk less obscured with black pigment–cells than in the case of the frog; and its gastrulation has better retained the original palingenetic character. It was first described by Scott and Osborn (1879), and Oscar Hertwig especially made a careful study of it (1881), and rightly pointed out its great importance in helping us to understand the vertebrate development. Its globular blastula (Figure 1.45) consists of loosely–aggregated, yelk–filled entodermic cells or yelk–cells (dz) in the lower vegetal half; the upper, animal half encloses the hemispherical segmentation–cavity (fh), the curved roof of which is formed of two or three strata of small ectodermic cells. At the point where the latter pass into the former (at the equator of the globular vesicle) we have the border zone (rz). The folding which leads to the formation of the gastrula takes place at a spot in this border zone, the primitive mouth (Figure 1.46 u).

Unequal segmentation takes place in some of the cyclostoma and in the oldest fishes in just the same way as in most of the amphibia. Among the cyclostoma ("round–mouthed") the familiar lampreys are particularly interesting. In respect of organisation and development they are half–way between the acrania (lancelet) and the lowest real fishes (Selachii); hence I divided the group of the cyclostoma in 1886 from the real fishes with which they were formerly associated, and formed of them a special class of vertebrates. The ovum–segmentation in our common river–lamprey (Petromyzon fluviatilis) was described by Max Schultze in 1856, and afterwards by Scott (1882) and Goette (1890).

Unequal total segmentation follows the same lines in the oldest fishes, the selachii and ganoids, which are directly descended from the cyclostoma. The primitive fishes (Selachii), which we must regard as the ancestral group of the true fishes, were generally considered, until a short time ago, to be discoblastic. It was not until the beginning of the twentieth century that Bashford Dean made the important discovery in Japan that one of the oldest living fishes of the shark type (Cestracion japonicus) has the same total unequal segmentation as the amphiblastic plated fishes (ganoides).* (* Bashford Dean, Holoblastic Cleavage in the Egg of a Shark, Cestracion japonicus Macleay. Annotationes zoologicae japonenses, volume 4 Tokio 1901.) This is particularly interesting in connection with our subject, because the few remaining survivors of this division, which was so numerous in paleozoic times, exhibit three different types of gastrulation. The oldest and most conservative forms of the modern ganoids are the scaly sturgeons (Sturiones), plated fishes of great evolutionary importance, the eggs of which are eaten as caviar; their cleavage is not essentially different from that of the lampreys and the amphibia. On the other hand, the most modern of the plated fishes, the beautifully scaled bony pike of the North American rivers (Lepidosteus), approaches the osseous fishes, and is discoblastic like them. A third genus (Amia) is midway between the sturgeons and the latter.

(FIGURE 1.48. Sagittal section of the gastrula of the water–salamander (Triton). (From Hertwig.) Letters as in Figure 1.47; except—p yelk–stopper, mk beginning of the middle germinal layer.)

The group of the lung–fishes (Dipneusta or Dipnoi) is closely connected with the older ganoids. In respect of their whole organisation they are midway between the gill–breathing fishes and the lung–breathing amphibia; they share with the former the shape of the body and limbs, and with the latter the form of the heart and lungs. Of the older dipnoi (Paladipneusta) we have now only one specimen, the remarkable Ceratodus of East Australia; its amphiblastic gastrulation has been recently explained by Richard Semon (cf. Chapter 2.21). That of the two modern dipneusta, of which Protopterus is found in Africa and Lepidosiren in America, is not materially different. (Cf. Figure 1.51.)

(FIGURE 1.49. Ovum–segmentation of the lamprey (Petromyzon fluviatalis), in four successive stages. The small cells of the upper (animal) hemisphere divide much more quickly than the cells of the lower (vegetal) hemisphere.

FIGURE 1.50. Gastrulation of the lamprey (Petromyzon fluviatilis). A blastula, with wide embryonic cavity (blastocoel, bl), g incipient invagination. B depula, with advanced invagination, from the primitive mouth (g). C gastrula, with complete primitive gut: the embryonic cavity has almost disappeared in consequence of invagination.)

All these amphiblastic vertebrates, Petromyzon and Cestracion, Accipenser and Ceratodus, and also the salamanders and batrachia, belong to the old, conservative groups of our stem. Their unequal ovum–segmentation and gastrulation have many peculiarities in detail, but can always be reduced with comparative ease to the original cleavage and gastrulation of the lowest vertebrate, the amphioxus; and this is little removed, as we have seen, from the very simple archigastrula of the Sagitta and Monoxenia (see Figures 1.29 to 1.36). All these and many other classes of animals generally agree in the circumstance that in segmentation their ovum divides into a large number of cells by repeated cleavage. All such ova have been called, after Remak, "whole–cleaving" (holoblasta), because their division into cells is complete or total.

(FIGURE 1.51. Gastrulation of ceratodus (from Semon). A and C stage with four cells, B and D with sixteen cells. A and B are seen from above, C and D sideways. E stage with thirty–two cells; F blastula; G gastrula in longitudinal section. fh segmentation–cavity. gh primitive gut or gastric cavity.)

In a great many other classes of animals this is not the case, as we find (in the vertebrate stem) among the birds, reptiles, and most of the fishes; among the insects and most of the spiders and crabs (of the articulates); and the cephalopods (of the molluscs). In all these animals the mature ovum, and the stem–cell that arises from it in fertilisation, consist of two different and separate parts, which we have called formative yelk and nutritive yelk. The formative yelk alone consists of living protoplasm, and is the active, evolutionary, and nucleated part of the ovum; this alone divides in segmentation, and produces the numerous cells which make up the embryo. On the other hand, the nutritive yelk is merely a passive part of the contents of the ovum, a subordinate element which contains nutritive material (albumin, fat, etc.), and so represents in a sense the provision–store of the developing embryo. The latter takes a quantity of food out of this store, and finally consumes it all. Hence the nutritive yelk is of great indirect importance in embryonic development, though it has no direct share in it. It either does not divide at all, or only later on, and does not generally consist of cells. It is sometimes large and sometimes small, but generally many times larger than the formative yelk; and hence it is that it was formerly thought the more important of the two. As the respective significance of these two parts of the ovum is often wrongly described, it must be borne in mind that the nutritive yelk is only a secondary addition to the primary cell, it is an inner enclosure, not an external appendage. All ova that have this independent nutritive yelk are called, after Remak, "partially–cleaving" (meroblasta). Their segmentation is incomplete or partial.

(FIGURE 1.52. Ovum of a deep–sea bony fish. b protoplasm of the stem–cell, k nucleus of same, d clear globule of albumin, the nutritive yelk, f fat–globule of same, c outer membrane of the ovum, or ovolemma.)

There are many difficulties in the way of understanding this partial segmentation and the gastrula that arises from it. We have only recently succeeded, by means of comparative research, in overcoming these difficulties, and reducing this cenogenetic form of gastrulation to the original palingenetic type. This is comparatively easy in the small meroblastic ova which contain little nutritive yelk—for instance, in the marine ova of a bony fish, the development of which I observed in 1875 at Ajaccio in Corsica. I found them joined together in lumps of jelly, floating on the surface of the sea; and, as the little ovula were completely transparent, I could easily follow the development of the germ step by step. These ovula are glossy and colourless globules of little more than the 50th of an inch. Inside a structureless, thin, but firm membrane (ovolemma, Figure 1.52 c) we find a large, quite clear, and transparent globule of albumin (d). At both poles of its axis this globule has a pit–like depression. In the pit at the upper, animal pole (which is turned downwards in the floating ovum) there is a bi–convex lens composed of protoplasm, and this encloses the nucleus (k); this is the formative yelk of the stem–cell, or the germinal disk (b). The small fat–globule (f) and the large albumin–globule (d) together form the nutritive yelk. Only the formative yelk undergoes cleavage, the nutritive yelk not dividing at all at first.

The segmentation of the lens–shaped formative yelk (b) proceeds quite independently of the nutritive yelk, and in perfect geometrical order.

When the mulberry–like cluster of cells has been formed, the border–cells of the lens separate from the rest and travel into the yelk and the border–layer. From this the blastula is developed; the regular bi–convex lens being converted into a disk, like a watch–glass, with thick borders. This lies on the upper and less curved polar surface of the nutritive yelk like the watch glass on the yelk. Fluid gathers between the outer layer and the border, and the segmentation–cavity is formed. The gastrula is then formed by invagination, or a kind of turning–up of the edge of the blastoderm. In this process the segmentation–cavity disappears.

The space underneath the entoderm corresponds to the primitive gut–cavity, and is filled with the decreasing food–yelk (n). Thus the formation of the gastrula of our fish is complete. In contrast to the two chief forms of gastrula we considered previously, we give the name of discoid gastrula (discogastrula, Figure 1.54) to this third principal type.

Very similar to the discoid gastrulation of the bony fishes is that of the hags or myxinoida, the remarkable cyclostomes that live parasitically in the body–cavity of fishes, and are distinguished by several notable peculiarities from their nearest relatives, the lampreys. While the amphiblastic ova of the latter are small and develop like those of the amphibia, the cucumber–shaped ova of the hag are about an inch long, and form a discoid gastrula. Up to the present it has only been observed in one species (Bdellostoma Stouti), by Dean and Doflein (1898).

It is clear that the important features which distinguish the discoid gastrula from the other chief forms we have considered are determined by the large food–yelk. This takes no direct part in the building of the germinal layers, and completely fills the primitive gut–cavity of the gastrula, even protruding at the mouth–opening. If we imagine the original bell–gastrula (Figures 1.30 to 1.36) trying to swallow a ball of food which is much bigger than itself, it would spread out round it in discoid shape in the attempt, just

127

as we find to be the case here (Figure 1.54). Hence we may derive the discoid gastrula from the original bell–gastrula, through the intermediate stage of the hooded gastrula. It has arisen through the accumulation of a store of food–stuff at the vegetal pole, a "nutritive yelk" being thus formed in contrast to the "formative yelk." Nevertheless, the gastrula is formed here, as in the previous cases, by the folding or invagination of the blastula. We can, therefore, reduce this cenogenetic form of the discoid segmentation to the palingenetic form of the primitive cleavage.

(FIGURE 1.53. Ovum–segmentation of a bony fish. A first cleavage of the stem–cell (cytula), B division of same into four segmentation–cells (only two visible), C the germinal disk divides into the blastoderm (b) and the periblast (p). d nutritive yelk, f fat–globule, c ovolemma, z space between the ovolemma and the ovum, filled with a clear fluid.)

This reduction is tolerably easy and confident in the case of the small ovum of our deep–sea bony fish, but it becomes difficult and uncertain in the case of the large ova that we find in the majority of the other fishes and in all the reptiles and birds. In these cases the food–yelk is, in the first place, comparatively colossal, the formative yelk being almost invisible beside it; and, in the second place, the food–yelk contains a quantity of different elements, which are known as "yelk–granules, yelk–globules, yelk–plates, yelk–flakes, yelk–vesicles," and so on. Frequently these definite elements in the yelk have been described as real cells, and it has been wrongly stated that a portion of the embryonic body is built up from these cells. This is by no means the case. In every case, however large it is—and even when cell–nuclei travel into it during the cleavage of the border—the nutritive yelk remains a dead accumulation of food, which is taken into the gut during embryonic development and consumed by the embryo. The latter develops solely from the living formative yelk of the stem–cell. This is equally true of the ova of our small bony fishes and of the colossal ova of the primitive fishes, reptiles, and birds.

(FIGURE 1.54. Discoid gastrula (discogastrula) of a bony fish. e ectoderm, i entoderm, w border–swelling or primitive mouth, n albuminous globule of the nutritive yelk, f fat–globule of same, c external membrane (ovolemma), d partition between entoderm and ectoderm (earlier the segmentation–cavity).)

The gastrulation of the primitive fishes or selachii (sharks and rays) has been carefully studied of late years by Ruckert, Rabl, and H.E. Ziegler in particular, and is very important in the sense that this group is the oldest among living fishes, and their gastrulation can be derived directly from that of the cyclostoma by the accumulation of a large quantity of food-yelk. The oldest sharks (Cestracion) still have the unequal segmentation inherited from the cyclostoma. But while in this case, as in the case of the amphibia, the small ovum completely divides into cells in segmentation, this is no longer so in the great majority of the selachii (or Elasmobranchii). In these the contractility of the active protoplasm no longer suffices to break up the huge mass of the passive deutoplasm completely into cells; this is only possible in the upper or dorsal part, but not in the lower or ventral section. Hence we find in the primitive fishes a blastula with a small eccentric segmentation-cavity (Figure 1.55 b), the wall of which varies greatly in composition. The circular border of the germinal disk which connects the roof and floor of the segmentation-cavity corresponds to the border-zone at the equator of the amphibian ovum. In the middle of its hinder border we have the beginning of the invagination of the primitive gut (Figure 1.56 ud); it extends gradually from this spot (which corresponds to the Rusconian anus of the amphibia) forward and around, so that the primitive mouth becomes first crescent-shaped and then circular, and, as it opens wider, surrounds the ball of the larger food-yelk.

Essentially different from the wide-mouthed discoid gastrula of most of the selachii is the narrow-mouthed discoid gastrula (or epigastrula) of the amniotes, the reptiles, birds, and monotremes; between the two—as an intermediate stage—we have the amphigastrula of the amphibia. The latter has developed from the amphigastrula of the ganoids and dipneusts, whereas the discoid amniote gastrula has been evolved from the amphibian gastrula by the addition of food-yelk. This change of gastrulation is still found in the remarkable ophidia (Gymnophiona, Coecilia, or Peromela), serpent-like amphibia that live in moist soil in the tropics, and in many respects represent the transition from the gill-breathing amphibia to the lung-breathing reptiles. Their embryonic development has been explained by the fine studies of the brothers Sarasin of Ichthyophis glutinosa at Ceylon (1887), and those of August Brauer of the Hypogeophis rostrata in the Seychelles (1897). It is only by the historical and comparative study of these that we can understand the difficult and obscure gastrulation of the amniotes.

The bird's egg is particularly important for our purpose, because most of the chief studies of the development of the vertebrates are based on observations of the hen's egg during hatching. The mammal ovum is much more difficult to obtain and study, and for this practical and obvious reason very rarely thoroughly investigated. But we can get hens' eggs in any quantity at any time, and, by means of artificial incubation, follow the development of the embryo step by step. The bird's egg differs considerably from the tiny mammal ovum in size, a large quantity of food–yelk accumulating within the original yelk or the protoplasm of the ovum. This is the yellow ball which we commonly call the yolk of the egg. In order to understand the bird's egg aright—for it is very often quite wrongly explained—we must examine it in its original condition, and follow it from the very beginning of its development in the bird's ovary. We then see that the original ovum is a quite small, naked, and simple cell with a nucleus, not differing in either size or shape from the original ovum of the mammals and other animals (cf. Figure 1.13 E). As in the case of all the craniota (animals with a skull), the original or primitive ovum (protovum) is covered with a continuous layer of small cells. This membrane is the follicle, from which the ovum afterwards issues. Immediately underneath it the structureless yelk–membrane is secreted from the yelk.

(FIGURE 1.55. Longitudinal section through the blastula of a shark (Pristiuris). (From Ruckert.) (Looked at from the left; to the right is the hinder end, H, to the left the fore end, V.) B segmentation–cavity, kz cells of the germinal membrane, dk yelk–nuclei.

FIGURE 1.56. Longitudinal section of the blastula of a shark (Pristiurus) at the beginning of gastrulation. (From Ruckert.) (Seen from the left.) V fore end, H hind end, B segmentation–cavity, ud first trace of the primitive gut, dk yelk–nuclei, fd fine–grained yelk, gd coarse–grained yelk.)

The small primitive ovum of the bird begins very early to take up into itself a quantity of food–stuff through the yelk–membrane, and work it up into the "yellow yelk." In this way the ovum enters on its second stage (the metovum), which is many times larger than the first, but still only a single enlarged cell. Through the accumulation of the store of yellow yelk within the ball of protoplasm the nucleus it contains (the germinal vesicle) is forced to the surface of the ball. Here it is surrounded by a small quantity of protoplasm, and with this forms the lens–shaped formative yelk (Figure 1.15 b). This is seen on the yellow yelk–ball, at a certain point of the surface, as a small round white spot—the "tread" (cicatricula). From this point a thread–like column of white nutritive yelk (d), which contains no yellow yelk–granules, and is softer than the yellow food–yelk, proceeds to the middle of the yellow yelk–ball, and forms there a small central globule of white yelk (Figure 1.15 d). The whole of this white yelk is not sharply separated from the yellow yelk, which shows a slight trace of concentric layers in the hard–boiled egg

(Figure 1.15 c). We also find in the hen's egg, when we break the shell and take out the yelk, a round small white disk at its surface which corresponds to the tread. But this small white "germinal disk" is now further developed, and is really the gastrula of the chick. The body of the chick is formed from it alone. The whole white and yellow yelk—mass is without any significance for the formation of the embryo, it being merely used as food by the developing chick. The clear, glarous mass of albumin that surrounds the yellow yelk of the bird's egg, and also the hard chalky shell, are only formed within the oviduct round the impregnated ovum.

When the fertilisation of the bird's ovum has taken place within the mother's body, we find in the lens—shaped stem—cell the progress of flat, discoid segmentation (Figure 1.57). First two equal segmentation—cells (A) are formed from the ovum. These divide into four (B), then into eight, sixteen (C), thirty—two, sixty—four, and so on. The cleavage of the cells is always preceded by a division of their nuclei. The cleavage surfaces between the segmentation—cells appear at the free surface of the tread as clefts. The first two divisions are vertical to each other, in the form of a cross (B). Then there are two more divisions, which cut the former at an angle of forty—five degrees. The tread, which thus becomes the germinal disk, now has the appearance of an eight—rayed star. A circular cleavage next taking place round the middle, the eight triangular cells divide into sixteen, of which eight are in the middle and eight distributed around (C). Afterwards circular clefts and radial clefts, directed towards the centre, alternate more or less irregularly (D, E). In most of the amniotes the formation of concentric and radial clefts is irregular from the very first; and so also in the hen's egg. But the final outcome of the cleavage—process is once more the formation of a large number of small cells of a similar nature. As in the case of the fish—ovum, these segmentation—cells form a round, lens—shaped disk, which corresponds to the morula, and is embedded in a small depression of the white yelk. Between the lens—shaped disk of the morula—cells and the underlying white yelk a small cavity is now formed by the accumulation of fluid, as in the fishes. Thus we get the peculiar and not easily recognisable blastula of the bird (Figure 1.58). The small segmentation—cavity (fh) is very flat and much compressed. The upper or dorsal wall (dw) is formed of a single layer of clear, distinctly separated cells; this corresponds to the upper or animal hemisphere of the triton—blastula (Figure 1.45). The lower or ventral wall of the flat dividing space (vw) is made up of larger and darker segmentation—cells; it corresponds to the lower or vegetal hemisphere of the blastula of the water—salamander (Figure 1.45 dz). The nuclei of the yelk—cells, which are in this case especially numerous at the edge of the lens—shaped blastula, travel into the white yelk, increase by cleavage, and contribute even to the further growth of the germinal disk by furnishing it with food–stuff.

(FIGURE 1.57. Diagram of discoid segmentation in the bird's ovum (magnified about ten times). Only the formative yelk (the tread) is shown in these six figures (A to F), because cleavage only takes place in this. The much larger food–yelk, which does not share in the cleavage, is left out and merely indicated by the dark ring without.)

The invagination or the folding inwards of the bird–blastula takes place in this case also at the hinder pole of the subsequent chief axis, in the middle of the hind border of the round germinal disk (Figure 1.59 s). At this spot we have the most brisk cleavage of the cells; hence the cells are more numerous and smaller here than in the fore–half of the germinal disk. The border–swelling or thick edge of the disk is less clear but whiter behind, and is more sharply separated from contiguous parts. In the middle of its hind border there is a white, crescent–shaped groove—Koller's sickle–groove (Fig 1.59 s); a small projecting process in the centre of it is called the sickle–knob (sk). This important cleft is the primitive mouth, which was described for a long time as the "primitive groove." If we make a vertical section through this part, we see that a flat and broad cleft stretches under the germinal disk forwards from the primitive mouth; this is the primitive gut (Figure 1.60 ud). Its roof or dorsal wall is formed by the folded upper part of the blastula, and its floor or ventral wall by the white yelk (wd), in which a number of yelk–nuclei (dk) are distributed. There is a brisk multiplication of these at the edge of the germinal disk, especially in the neighbourhood of the sickle–shaped primitive mouth.

We learn from sections through later stages of this discoid bird–gastrula that the primitive gut–cavity, extending forward from the primitive mouth as a flat pouch, undermines the whole region of the round flat lens–shaped blastula (Figure 1.61 ud). At the same time, the segmentation–cavity gradually disappears altogether, the folded inner germinal layer (ik) placing itself from underneath on the overlying outer germinal layer (ak). The typical process of invagination, though greatly disguised, can thus be clearly seen in this case, as Goette and Rauber, and more recently Duval (Figure 1.61), have shown.

(FIGURE 1.58. Vertical section of the blastula of a hen (discoblastula). fh segmentation–cavity, dw dorsal wall of same, vw ventral wall, passing directly into the white yelk (wd) (From Duval.)

FIGURE 1.59. The germinal disk of the hen's ovum at the beginning of gastrulation; A before incubation, B in the first hour of incubation. (From Koller.) ks germinal–disk, V its fore and H its hind border; es embryonic shield, s sickle–groove, sk sickle knob, d yelk.

FIGURE 1.60. Longitudinal section of the germinal disk of a siskin (discogastrula). (From Duval.) ud primitive gut, vl, hl fore and hind lips of the primitive mouth (or sickle–edge); ak outer germinal layer, ik inner germinal layer, dk yelk–nuclei, wd white yelk.

FIGURE 1.61. Longitudinal section of the discoid gastrula of the nightingale. (From Duval.) ud primitive gut, vl, hl fore and hind lips of the primitive mouth; ak, ik outer and inner germinal layers; vr fore–border of the discogastrula.)

The older embryologists (Pander, Baer, Remak), and, in recent times especially, His, Kolliker, and others, said that the two primary germinal layers of the hen's ovum—the oldest and most frequent subject of observation!—arose by horizontal cleavage of a simple germinal disk. In opposition to this accepted view, I affirmed in my Gastraea Theory (1873) that the discoid bird–gastrula, like that of all other vertebrates, is formed by folding (or invagination), and that this typical process is merely altered in a peculiar way and disguised by the immense accumulation of food–yelk and the flat spreading of the discoid blastula at one part of its surface. I endeavoured to establish this view by the derivation of the vertebrates from one source, and especially by proving that the birds descend from the reptiles, and these from the amphibia. If this is correct, the discoid gastrula of the amniotes must have been formed by the folding–in of a hollow blastula, as has been shown by Remak and Rusconi of the discoid gastrula of the amphibia, their direct ancestors. The accurate and extremely careful observations of the authors I have mentioned (Goette, Rauber, and Duval) have decisively proved this recently for the birds; and the same has been done for the reptiles by the fine studies of Kupffer, Beneke, Wenkebach, and others. In the shield–shaped germinal disk of the lizard (Figure 1.62), the crocodile, the tortoise, and other reptiles, we find in the middle of the hind border (at the same spot as the sickle groove in the bird) a transverse furrow (u), which leads into a flat, pouch–like, blind sac, the primitive gut. The fore (dorsal) and hind (ventral) lips of the transverse furrow correspond exactly to the lips of the primitive mouth (or sickle–groove) in the birds.

133

(FIGURE 1.62. Germinal disk of the lizard (Lacerta agilis). (From Kupffer.) u primitive mouth, s sickle, es embryonic shield, hf and df light and dark germinative area.)

The gastrulation of the mammals must be derived from this special embryonic development of the reptiles and birds. This latest and most advanced class of the vertebrates has, as we shall see afterwards, evolved at a comparatively recent date from an older group of reptiles; and all these amniotes must have come originally from a common stem–form. Hence the distinctive embryonic process of the mammal must have arisen by cenogenetic modifications from the older form of gastrulation of the reptiles and birds. Until we admit this thesis we cannot understand the formation of the germinal layers in the mammal, and therefore in man.

I first advanced this fundamental principle in my essay On the Gastrulation of Mammals (1877), and sought to show in this way that I assumed a gradual degeneration of the food–yelk and the yelk–sac on the way from the proreptiles to the mammals. "The cenogenetic process of adaptation," I said, "which has occasioned the atrophy of the rudimentary yelk–sac of the mammal, is perfectly clear. It is due to the fact that the young of the mammal, whose ancestors were certainly oviparous, now remain a long time in the womb. As the great store of food–yelk, which the oviparous ancestors gave to the egg, became superfluous in their descendants owing to the long carrying in the womb, and the maternal blood in the wall of the uterus made itself the chief source of nourishment, the now useless yelk–sac was bound to atrophy by embryonic adaptation."

My opinion met with little approval at the time; it was vehemently attacked by Kolliker, Hensen, and His in particular. However, it has been gradually accepted, and has recently been firmly established by a large number of excellent studies of mammal gastrulation, especially by Edward Van Beneden's studies of the rabbit and bat, Selenka's on the marsupials and rodents, Heape's and Lieberkuhn's on the mole, Kupffer and Keibel's on the rodents, Bonnet's on the ruminants, etc. From the general comparative point of view, Carl Rabl in his theory of the mesoderm, Oscar Hertwig in the latest edition of his Manual (1902), and Hubrecht in his Studies in Mammalian Embryology (1891), have supported the opinion, and sought to derive the peculiarly modified gastrulation of the mammal from that of the reptile.

(FIGURE 1.63. Ovum of the opossum (Didelphys) divided into four. (From Selenka.) b the four segmentation–cells, r directive body, c unnucleated coagulated matter, p, albumin–membrane.)

In the meantime (1884) the studies of Wilhelm Haacke and Caldwell provided a proof of the long–suspected and very interesting fact, that the lowest mammals, the monotremes, LAY EGGS, like the birds and reptiles, and are not viviparous like the other mammals. Although the gastrulation of the monotremes was not really known until studied by Richard Semon in 1894, there could be little doubt, in view of the great size of their food–yelk, that their ovum–segmentation was discoid, and led to the formation of a sickle–mouthed discogastrula, as in the case of the reptiles and birds. Hence I had, in 1875 (in my essay on The Gastrula and Ovum–segmentation of Animals), counted the monotremes among the discoblastic vertebrates. This hypothesis was established as a fact nineteen years afterwards by the careful observations of Semon; he gave in the second volume of his great work, Zoological Journeys in Australia (1894), the first description and correct explanation of the discoid gastrulation of the monotremes. The fertilised ova of the two living monotremes (Echidna and Ornithorhynchus) are balls of one–fifth of an inch in diameter, enclosed in a stiff shell; but they grow considerably during development, so that when laid the egg is three times as large. The structure of the plentiful yelk, and especially the relation of the yellow and the white yelk, are just the same as in the reptiles and birds. As with these, partial cleavage takes place at a spot on the surface at which the small formative yelk and the nucleus it encloses are found. First is formed a lens–shaped circular germinal disk. This is made up of several strata of cells, but it spreads over the yelk–ball, and thus becomes a one–layered blastula. If we then imagine the yelk it contains to be dissolved and replaced by a clear liquid, we have the characteristic blastula of the higher mammals. In these the gastrulation proceeds in two phases, as Semon rightly observes: firstly, formation of the entoderm by cleavage at the centre and further growth at the edge; secondly, invagination. In the monotremes more primitive conditions have been retained better than in the reptiles and birds. In the latter, before the commencement of the gastrula–folding, we have, at least at the periphery, a two–layered embryo forming from the cleavage. But in the monotremes the formation of the cenogenetic entoderm does not precede the invagination; hence in this case the construction of the germinal layers is less modified than in the other amniota.

The marsupials, a second sub–class, come next to the oviparous monotremes, the oldest of the mammals. But as in their case the food–yelk is already atrophied, and the little ovum develops within the mother's body, the partial cleavage has been reconverted into total. One section of the marsupials still show points of agreement with the monotremes, while another section of them, according to the splendid investigations of Selenka, form a connecting–link between these and the placentals.

(FIGURE 1.64. Blastula of the opossum (Didelphys). (From Selenka.) a animal pole of the blastula, v vegetal pole, en mother–cell of the entoderm, ex ectodermic cells, s spermia, ib unnucleated yelk–balls (remainder of the food–yelk), p albumin membrane.)

The fertilised ovum of the opossum (Didelphys) divides, according to Selenka, first into two, then four, then eight equal cells; hence the segmentation is at first equal or homogeneous. But in the course of the cleavage a larger cell, distinguished by its less clear plasm and its containing more yelk–granules (the mother cell of the entoderm, Figure 1.64 en), separates from the others; the latter multiply more rapidly than the former. As, further, a quantity of fluid gathers in the morula, we get a round blastula, the wall of which is of varying thickness, like that of the amphioxus (Figure 1.38 E) and the amphibia (Figure 1.45). The upper or animal hemisphere is formed of a large number of small cells; the lower or vegetal hemisphere of a small number of large cells. One of the latter, distinguished by its size (Figure 1.64 en), lies at the vegetal pole of the blastula–axis, at the point where the primitive mouth afterwards appears. This is the mother–cell of the entoderm; it now begins to multiply by cleavage, and the daughter–cells (Figure 1.65 i) spread out from this spot over the inner surface of the blastula, though at first only over the vegetal hemisphere. The less clear entodermic cells (i) are distinguished at first by their rounder shape and darker nuclei from the higher, clearer, and longer entodermic cells (e), afterwards both are greatly flattened, the inner blastodermic cells more than the outer.

(FIGURE 1.65. Blastula of the opossum (Didelphys) at the beginning of gastrulation. (From Selenka.) e ectoderm, i entoderm; a animal pole, u primitive mouth at the vegetal pole, f segmentation–cavity, d unnucleated yelk–balls (relics of the reduced food–yelk), c nucleated curd (without yelk–granules).

FIGURE 1.66. Oval gastrula of the opossum (Didelphys), about eight hours old. (From Selenka) (external view).)

The unnucleated yelk–balls and curd (Figure 1.65 d) that we find in the fluid of the blastula in these marsupials are very remarkable; they are the relics of the atrophied food–yelk, which was developed in their ancestors, the monotremes, and in the reptiles.

136

In the further course of the gastrulation of the opossum the oval shape of the gastrula (Figure 1.66) gradually changes into globular, a larger quantity of fluid accumulating in the vesicle. At the same time, the entoderm spreads further and further over the inner surface of the ectoderm (e). A globular vesicle is formed, the wall of which consists of two thin simple strata of cells; the cells of the outer germinal layer are rounder, and those of the inner layer flatter. In the region of the primitive mouth (p) the cells are less flattened, and multiply briskly. From this point—from the hind (ventral) lip of the primitive mouth, which extends in a central cleft, the primitive groove—the construction of the mesoderm proceeds.

Gastrulation is still more modified and curtailed cenogenetically in the placentals than in the marsupials. It was first accurately known to us by the distinguished investigations of Edward Van Beneden in 1875, the first object of study being the ovum of the rabbit. But as man also belongs to this sub–class, and as his as yet unstudied gastrulation cannot be materially different from that of the other placentals, it merits the closest attention. We have, in the first place, the peculiar feature that the two first segmentation–cells that proceed from the cleavage of the fertilised ovum (Figure 1.68) are of different sizes and natures; the difference is sometimes greater, sometimes less (Figure 1.69). One of these first daughter–cells of the ovum is a little larger, clearer, and more transparent than the other. Further, the smaller cell takes a colour in carmine, osmium, etc., more strongly than the larger. By repeated cleavage of it a morula is formed, and from this a blastula, which changes in a very characteristic way into the greatly modified gastrula. When the number of the segmentation–cells in the mammal embryo has reached ninety–six (in the rabbit, about seventy hours after impregnation) the foetus assumes a form very like the archigastrula (Figure 1.72). The spherical embryo consists of a central mass of thirty–two soft, round cells with dark nuclei, which are flattened into polygonal shape by mutual pressure, and colour dark–brown with osmic acid (Figure 1.72 i). This dark central group of cells is surrounded by a lighter spherical membrane, consisting of sixty–four cube–shaped, small, and fine–grained cells which lie close together in a single stratum, and only colour slightly in osmic acid (Figure 1.72 e). The authors who regard this embryonic form as the primary gastrula of the placental conceive the outer layer as the ectoderm and the inner as the entoderm. The entodermic membrane is only interrupted at one spot, one, two, or three of the ectodermic cells being loose there. These form the yelk–stopper, and fill up the mouth of the gastrula (a). The central primitive gut–cavity (d) is full of entodermic cells. The uni–axial type of the mammal gastrula is accentuated in this way. However, opinions still differ considerably as to the real nature of this "provisional gastrula" of the placental and its relation to the blastula into which it is converted.

As the gastrulation proceeds a large spherical blastula is formed from this peculiar solid amphigastrula of the placental, as we saw in the case of the marsupial. The accumulation of fluid in the solid gastrula (Figure 1.73 A) leads to the formation of an eccentric cavity, the group of the darker entodermic cells (hy) remaining directly attached at one spot with the round enveloping stratum of the lighter ectodermic cells (ep). This spot corresponds to the original primitive mouth (prostoma or blastoporus). From this important spot the inner germinal layer spreads all round on the inner surface of the outer layer, the cell–stratum of which forms the wall of the hollow sphere; the extension proceeds from the vegetal towards the animal pole.

(FIGURE 1.67. Longitudinal section through the oval gastrula of the opossum (Figure 1.69). (From Selenka.) p primitive mouth, e ectoderm, i entoderm, d yelk remains in the primitive gut–cavity (u).)

The cenogenetic gastrulation of the placental has been greatly modified by secondary adaptation in the various groups of this most advanced and youngest sub–class of the mammals. Thus, for instance, we find in many of the rodents (guinea–pigs, mice, etc.) APPARENTLY a temporary inversion of the two germinal layers. This is due to a folding of the blastodermic wall by what is called the "girder," a plug–shaped growth of Rauber's "roof–layer." It is a thin layer of flat epithelial cells, that is freed from the surface of the blastoderm in some of the rodents; it has no more significance in connection with the general course of placental gastrulation than the conspicuous departure from the usual globular shape in the blastula of some of the ungulates. In some pigs and ruminants it grows into a thread–like, long and thin tube.

(FIGURE 1.68. Stem–cell of the mammal ovum (from the rabbit). k stem–nucleus, n nuclear corpuscle, p protoplasm of the stem–cell, z modified zona pellucida, h outer albuminous membrane, s dead sperm–cells.

FIGURE 1.69. Incipient cleavage of the mammal ovum (from the rabbit). The stem–cell has divided into two unequal cells, one lighter (e) and one darker (i). z zona pellucida, h outer albuminous membrane, s dead sperm–cell.

FIGURE 1.70. The first four segmentation–cells of the mammal ovum (from the rabbit). e the two larger (and lighter) cells, i the two smaller (and darker) cells, z zona pellucida, h outer albuminous membrane.

FIGURE 1.71. Mammal ovum with eight segmentation–cells (from the rabbit). e four larger and lighter cells, i four smaller and darker cells, z zona pellucida, h outer albuminous membrane.)

Thus the gastrulation of the placentals, which diverges most from that of the amphioxus, the primitive form, is reduced to the original type, the invagination of a modified blastula. Its chief peculiarity is that the folded part of the blastoderm does not form a completely closed (only open at the primitive mouth) blind sac, as is usual; but this blind sac has a wide opening at the ventral curve (opposite to the dorsal mouth); and through this opening the primitive gut communicates from the first with the embryonic cavity of the blastula. The folded crest–shaped entoderm grows with a free circular border on the inner surface of the entoderm towards the vegetal pole; when it has reached this, and the inner surface of the blastula is completely grown over, the primitive gut is closed. This remarkable direct transition of the primitive gut–cavity into the segmentation–cavity is explained simply by the assumption that in most of the mammals the yelk–mass, which is still possessed by the oldest forms of the class (the monotremes) and their ancestors (the reptiles), is atrophied. This proves the essential unity of gastrulation in all the vertebrates, in spite of the striking differences in the various classes.

In order to complete our consideration of the important processes of segmentation and gastrulation, we will, in conclusion, cast a brief glance at the fourth chief type—superficial segmentation. In the vertebrates this form is not found at all. But it plays the chief part in the large stem of the articulates—the insects, spiders, myriapods, and crabs. The distinctive form of gastrula that comes of it is the "vesicular gastrula" (Perigastrula).

In the ova which undergo this superficial cleavage the formative yelk is sharply divided from the nutritive yelk, as in the preceding cases of the ova of birds, reptiles, fishes, etc.; the formative yelk alone undergoes cleavage. But while in the ova with discoid gastrulation the formative yelk is not in the centre, but at one pole of the uni–axial ovum, and the food–yelk gathered at the other pole, in the ova with superficial cleavage we find the formative yelk spread over the whole surface of the ovum; it encloses spherically the food–yelk, which is accumulated in the middle of the ova. As the segmentation only affects the former and not the latter, it is bound to be entirely "superficial"; the store of food in the middle is quite untouched by it. As a rule, it proceeds in regular geometrical progression. In the end the whole of the formative yelk divides into a number of small and homogeneous cells, which lie close together in a single stratum on the entire surface of the ovum, and form a superficial blastoderm. This blastoderm is a simple, completely closed vesicle, the internal cavity of which is entirely full of food–yelk. This real blastula only differs from that of the primitive ova in its chemical composition. In the latter the content is water or a watery jelly; in the former it is a thick mixture, rich in food–yelk, of albuminous and fatty substances. As this quantity of food–yelk fills the centre of the ovum before cleavage begins, there is no difference in this respect between the morula and the blastula. The two stages rather agree in this.

When the blastula is fully formed, we have again in this case the important folding or invagination that determines gastrulation. The space between the skin–layer and the gut–layer (the remainder of the segmentation–cavity) remains full of food–yelk, which is gradually used up. This is the only material difference between our vesicular gastrula (perigastrula) and the original form of the bell–gastrula (archigastrula). Clearly the one has been developed from the other in the course of time, owing to the accumulation of food–yelk in the centre of the ovum.* (* On the reduction of all forms of gastrulation to the original palingenetic form see especially the lucid treatment of the subject in Arnold Lang's Manual of Comparative Anatomy (1888), Part 1.)

We must count it an important advance that we are thus in a position to reduce all the various embryonic phenomena in the different groups of animals to these four principal forms of segmentation and gastrulation. Of these four forms we must regard one only as the original palingenetic, and the other three as cenogenetic and derivative. The unequal, the discoid, and the superficial segmentation have all clearly arisen by secondary adaptation from the primary segmentation; and the chief cause of their development has been the gradual formation of the food–yelk, and the increasing antithesis between animal and vegetal halves of the ovum, or between ectoderm (skin–layer) and entoderm (gut–layer).

(FIGURE 1.72. Gastrula of the placental mammal (epigastrula from the rabbit), longitudinal section through the axis. e ectodermic cells (sixty–four, lighter and smaller), i entodermic cells (thirty–two, darker and larger), d central entodermic cell, filling the primitive gut—cavity, o peripheral entodermic cell, stopping up the opening of the primitive mouth (yelk—stopper in the Rusconian anus).)

(FIGURE 1.73. Gastrula of the rabbit. A as a solid, spherical cluster of cells, B changing into the embryonic vesicle, bp primitive mouth, ep ectoderm, hy entoderm.)

The numbers of careful studies of animal gastrulation that have been made in the last few decades have completely established the views I have expounded, and which I first advanced in the years 1872 to 1876. For a time they were greatly disputed by many embryologists. Some said that the original embryonic form of the metazoa was not the gastrula, but the "planula"—a double—walled vesicle with closed cavity and without mouth—aperture; the latter was supposed to pierce through gradually. It was afterwards shown that this planula (found in several sponges, etc.) was a later evolution from the gastrula. It was also shown that what is called delamination—the rise of the two primary germinal layers by the folding of the surface of the blastoderm (for instance, in the Geryonidae and other medusae)—was a secondary formation, due to cenogenetic variations from the original invagination of the blastula. The same may be said of what is called "immigration," in which certain cells or groups of cells are detached from the simple layer of the blastoderm, and travel into the interior of the blastula; they attach themselves to the inner wall of the blastula, and form a second internal epithelial layer—that is to say, the entoderm. In these and many other controversies of modern embryology the first requisite for clear and natural explanation is a careful and discriminative distinction between palingenetic (hereditary) and cenogenetic (adaptive) processes. If this is properly attended to, we find evidence everywhere of the biogenetic law.

CHAPTER 1.10. THE COELOM THEORY.

The two "primary germinal layers" which the gastraea theory has shown to be the first foundation in the construction of the body are found in this simplest form throughout life only in animals of the lowest grade—in the gastraeads, olynthus (the stem—form of the sponges), hydra, and similar very simple animals. In all the other animals new strata of cells are formed subsequently between these two primary body-layers, and these are generally comprehended under the title of the middle layer, or mesoderm. As a rule, the various products of this middle layer afterwards constitute the great bulk of the animal frame, while the original entoderm, or internal germinal layer, is restricted to the clothing of the alimentary canal and its glandular appendages; and, on the other hand, the ectoderm, or external germinal layer, furnishes the outer clothing of the body, the skin and nervous system.

In some large groups of the lower animals, such as the sponges, corals, and flat–worms, the middle germinal layer remains a single connected mass, and most of the body is developed from it; these have been called the three–layered metazoa, in opposition to the two–layered animals described. Like the two–layered animals, they have no body–cavity—that is to say, no cavity distinct from the alimentary system. On the other hand, all the higher animals have this real body–cavity (coeloma), and so are called coelomaria. In all these we can distinguish four secondary germinal layers, which develop from the two primary layers. To the same class belong all true vermalia (excepting the platodes), and also the higher typical animal stems that have been evolved from them—molluscs, echinoderms, articulates, tunicates, and vertebrates.

(FIGURES 1.74 AND 1.75. Diagram of the four secondary germinal layers, transverse section through the metazoic embryo: Figure 1.74 of an annelid, Figure 1.75 of a vermalian. a primitive gut, dd ventral glandular layer, df ventral fibre–layer, hm skin–fibre–layer, hs skin–sense–layer, u beginning of the rudimentary kidneys, n beginning of the nerve–plates.)

The body–cavity (coeloma) is therefore a new acquisition of the animal body, much younger than the alimentary system, and of great importance. I first pointed out this fundamental significance of the coelom in my Monograph on the Sponges (1872), in the section which draws a distinction between the body–cavity and the gut–cavity, and which follows immediately on the germ–layer theory and the ancestral tree of the animal kingdom (the first sketch of the gastraea theory). Up to that time these two principal cavities of the animal body had been confused, or very imperfectly distinguished; chiefly because Leuckart, the founder of the coelenterata group (1848), has attributed a body–cavity, but not a gut–cavity, to these lowest metazoa. In reality, the truth is just the other way about.

The ventral cavity, the original organ of nutrition in the multicellular animal–body, is the oldest and most important organ of all the metazoa, and, together with the primitive mouth, is formed in every case in the gastrula as the primitive gut; it is only at a much later stage that the body–cavity, which is entirely wanting in the coelenterata, is developed in some of the metazoa between the ventral and the body wall. The two cavities are entirely different in content and purport. The alimentary cavity (enteron) serves the purpose of digestion; it contains water and food taken from without, as well as the pulp (chymus) formed from this by digestion. On the other hand, the body–cavity, quite distinct from the gut and closed externally, has nothing to do with digestion; it encloses the gut itself and its glandular appendages, and also contains the sexual products and a certain amount of blood or lymph, a fluid that is transuded through the ventral wall.

As soon as the body–cavity appears, the ventral wall is found to be separated from the enclosing body–wall, but the two continue to be directly connected at various points. We can also then always distinguish a number of different layers of tissue in both walls—at least two in each. These tissue–layers are formed originally from four different simple cell–layers, which are the much–discussed four secondary germinal layers. The outermost of these, the skin–sense–layer (Figures 1.74 and 1.75 hs), and the innermost, the gut–gland–layer (dd), remain at first simple epithelia or covering–layers. The one covers the outer surface of the body, the other the inner surface of the ventral wall; hence they are called confining or limiting layers. Between them are the two middle–layers, or mesoblasts, which enclose the body–cavity.

(FIGURE 1.76. Coelomula of sagitta (gastrula with a couple of coelom–pouches. (From Kowalevsky.) bl.p primitive mouth, al primitive gut, pv coelom–folds, m permanent mouth.)

The four secondary germinal layers are so distributed in the structure of the body in all the coelomaria (or all metazoa that have a body–cavity) that the outer two, joined fast together, constitute the body–wall, and the inner two the ventral wall; the two walls are separated by the cavity of the coelom. Each of the walls is made up of a limiting layer and a middle layer. The two limiting layers chiefly give rise to epithelia, or covering–tissues, and glands and nerves, while the middle layers form the great bulk of the fibrous tissue, muscles, and connective matter. Hence the latter have also been called fibrous or muscular layers. The outer middle layer, which lies on the inner side of the skin–sense–layer, is the skin fibre–layer; the inner middle layer, which attaches from without to the ventral glandular layer, is the ventral fibre layer. The former is usually called briefly the parietal, and the latter the visceral layer or mesoderm. Of the many different names that have been given to the four secondary germinal layers, the following are those most in use to–day:—

1. Skin–sense–layer (outer limiting layer) and 2. Skin–fibre–layer (outer middle layer).

I. Neural layer (neuroblast) and II. Parietal layer (myoblast). The two secondary germinal layers of the body–wall: 1. Epithelial. 2. Fibrous.

3. Gut–fibre–layer (inner middle layer) and 4. Gut–gland–layer (inner limiting layer).

III. Visceral layer (gonoblast) and IV. Enteral layer (enteroblast). The two secondary germinal layers of the gut–wall: 3. Fibrous. 4. Epithelial.

The first scientist to recognise and clearly distinguish the four secondary germinal layers was Baer. It is true that he was not quite clear as to their origin and further significance, and made several mistakes in detail in explaining them. But, on the whole, their great importance did not escape him. However, in later years his view had to be given up in consequence of more accurate observations. Remak then propounded a three–layer theory, which was generally accepted. These theories of cleavage, however, began to give way thirty years ago, when Kowalevsky (1871) showed that in the case of Sagitta (a very clear and typical subject of gastrulation) the two middle germinal layers and the two limiting layers arise not by cleavage, but by folding—by a secondary invagination of the primary inner germ–layer. This invagination or folding proceeds from the primitive mouth, at the two sides of which (right and left) a couple of pouches are formed. As these coelom–pouches or coelom–sacs detach themselves from the primitive gut, a double body–cavity is formed (Figures 1.74 to 1.76).

(FIGURE 1.77. Coelomula of sagitta, in section. (From Hertwig.) D dorsal side, V ventral side, ik inner germinal layer, mv visceral mesoblast, lh body–cavity, mp parietal mesoblast, ak outer germinal layer.)

The same kind of coelom–formation as in sagitta was afterwards found by Kowalevsky in brachiopods and other invertebrates, and in the lowest vertebrate—the amphioxus. Further instances were discovered by two English embryologists, to whom we owe very considerable advance in ontogeny—E. Ray–Lankester and F. Balfour. On the strength of these and other studies, as well as most extensive research of their own, the brothers Oscar and Richard Hertwig constructed in 1881 the Coelom Theory. In order to appreciate fully the great merit of this illuminating and helpful theory, one must remember what a chaos of contradictory views was then represented by the "problem of the mesoderm," or the much–disputed "question of the origin of the middle germinal layer." The coelom theory brought some light and order into this infinite confusion by establishing the following points: 1. The body–cavity originates in the great majority of animals (especially in all the vertebrates) in the same way as in sagitta: a couple of pouches or sacs are formed by folding inwards at the primitive mouth, between the two primary germinal layers; as these pouches detach from the primitive gut, a pair of coelom–sacs (right and left) are formed; the coalescence of these produces a simple body–cavity. 2. When these coelom–embryos develop, not as a pair of hollow pouches, but as solid layers of cells (in the shape of a pair of mesodermal streaks)—as happens in the higher vertebrates—we have a secondary (cenogenetic) modification of the primary (palingenetic) structure; the two walls of the pouches, inner and outer, have been pressed together by the expansion of the large food–yelk. 3. Hence the mesoderm consists from the first of TWO genetically distinct layers, which do not originate by the cleavage of a primary simple middle layer (as Remak supposed). 4. These two middle layers have, in all vertebrates, and the great majority of the invertebrates, the same radical significance for the construction of the animal body; the inner middle layer, or the visceral mesoderm, (gut–fibre layer), attaches itself to the original entoderm, and forms the fibrous, muscular, and connective part of the visceral wall; the outer middle layer, or the parietal mesoderm (skin–fibre–layer), attaches itself to the original ectoderm and forms the fibrous, muscular, and connective part of the body–wall. 5. It is only at the point of origination, the primitive mouth and its vicinity, that the four secondary germinal layers are directly connected; from this point the two middle layers advance forward separately between the two primary germinal layers, to which they severally attach themselves. 6. The further separation or differentiation of the four secondary germinal layers and their division into the various tissues and organs take place especially in the later fore–part or head of the embryo, and extend backwards from there towards the primitive mouth.

(FIGURE 1.78. Section of a young sagitta. (From Hertwig.) dh visceral cavity, ik and ak inner and outer limiting layers, mv and mp inner and outer middle layers, lk body–cavity, dm and vm dorsal and visceral mesentery.)

All animals in which the body–cavity demonstrably arises in this way from the primitive gut (vertebrates, tunicates, echinoderms, articulates, and a part of the vermalia) were comprised by the Hertwigs under the title of enterocoela, and were contrasted with the other groups of the pseudocoela (with false body–cavity) and the coelenterata (with no body–cavity). However, this radical distinction and the views as to classification which it occasioned have been shown to be untenable. Further, the absolute differences in tissue–formation which the Hertwigs set up between the enterocoela and pseudocoela cannot be sustained in this connection. For these and other reasons their coelom–theory has been much criticised and partly abandoned. Nevertheless, it has rendered a great and lasting service in the solution of the difficult problem of the mesoderm, and a material part of it will certainly be retained. I consider it an especial merit of the theory that it has established the identity of the development of the two middle layers in all the vertebrates, and has traced them as cenogenetic modifications back to the original palingenetic form of development that we still find in the amphioxus. Carl Rabl comes to the same conclusion in his able Theory of the Mesoderm, and so do Ray–Lankester, Rauber, Kupffer, Ruckert, Selenka, Hatschek, and others. There is a general agreement in these and many other recent writers that all the different forms of coelom–construction, like those of gastrulation, follow one and the same strict hereditary law in the vast vertebrate stem; in spite of their apparent differences, they are all only cenogenetic modifications of one palingenetic type, and this original type has been preserved for us down to the present day by the invaluable amphioxus.

(FIGURES 1.79 AND 1.80. Transverse section of amphioxus–larvae. (From Hatschek.) Figure 1.79 at the commencement of coelom formation (still without segments), Figure 1.80 at the stage with four primitive segments. ak, ik, mk outer, inner, and middle germinal layer, hp horn plate, mp medullary plate, ch chorda, asterisk and asterisk, disposition of the coelom–pouches, lh body–cavity.)

But before we go into the regular coelomation of the amphioxus, we will glance at that of the arrow–worm (Sagitta), a remarkable deep–sea worm that is interesting in many ways for comparative anatomy and ontogeny. On the one hand, the transparency of the body and the embryo, and, on the other hand, the typical simplicity of its embryonic development, make the sagitta a most instructive object in connection with various problems. The class of the chaetogatha, which is only represented by the cognate genera of Sagitta and Spadella, is in another respect also a most remarkable branch of the extensive vermalia stem. It was therefore very gratifying that Oscar Hertwig (1880) fully explained the anatomy, classification, and evolution of the chaetognatha in his careful monograph.

The spherical blastula that arises from the impregnated ovum of the sagitta is converted by a folding at one pole into a typical archigastrula, entirely similar to that of the Monoxenia which I described (Chapter 1.8, Figure 1.29). This oval, uni–axial cup–larva (circular in section) becomes bilateral (or tri–axial) by the growth of a couple of coelom–pouches from the primitive gut (Figures 1.76 and 1.77). To the right and left a sac–shaped fold appears towards the top pole (where the permanent mouth, m, afterwards arises). The two sacs are at first separated by a couple of folds of the entoderm (Figure 1.76 pv), and are still connected with the primitive gut by wide apertures; they also communicate for a short time with the dorsal side (Figure 1.77 d). Soon, however, the coelom–pouches completely separate from each other and from the primitive gut; at the same time they enlarge so much that they close round the primitive gut (Figure 1.78). But in the middle line of the dorsal and ventral sides the pouches remain separated, their approaching walls joining here to form a thin vertical partition, the mesentery (dm and vm). Thus Sagitta has throughout life a double body–cavity (Figure 1.78 lk), and the gut is fastened to the body–wall both above and below by a mesentery—below by the ventral mesentery (vm), and above by the dorsal mesentery (dm). The inner layer of the two coelom–pouches (mv) attaches itself to the entoderm (ik), and forms with it the visceral wall. The outer layer (mp) attaches itself to the ectoderm (ak), and forms with it the outer body–wall. Thus we have in Sagitta a perfectly clear and simple illustration of the original coelomation of the enterocoela. This palingenetic fact is the more important, as the greater part of the two body–cavities in Sagitta changes afterwards into sexual glands—the fore or female part into a pair of ovaries, and the hind or male part into a pair of testicles.

Coelomation takes place with equal clearness and transparency in the case of the amphioxus, the lowest vertebrate, and its nearest relatives, the invertebrate tunicates, the sea–squirts. However, in these two stems, which we class together as Chordonia, this important process is more complex, as two other processes are associated with it—the development of the chorda from the entoderm and the separation of the medullary plate or nervous centre from the ectoderm. Here again the skulless amphioxus has preserved to our own time by tenacious heredity the chief phenomena in their original form, while it has been more or less modified by embryonic adaptation in all the other vertebrates (with skulls). Hence we must once more thoroughly understand the palingenetic embryonic features of the lancelet before we go on to consider the cenogenetic forms of the craniota.

(FIGURES 1.81 AND 1.82. Transverse section of amphioxus embryo. Figure 1.81 at the stage with five somites, Figure 1.82 at the stage with eleven somites. (From Hatschek.) ak outer germinal layer, mp medullary plate, n nerve–tube, ik inner germinal layer, dh visceral cavity, lh body–cavity, mk middle germinal layer (mk1 parietal, mk2 visceral), us primitive segment, ch chorda.)

The coelomation of the amphioxus, which was first observed by Kowalevsky in 1867, has been very carefully studied since by Hatschek (1881). According to him, there are first formed on the bilateral gastrula we have already considered (Figures 1.36 and 1.37) three parallel longitudinal folds—one single ectodermal fold in the central line of the dorsal surface, and a pair of entodermic folds at the two sides of the former. The broad ectodermal fold that first appears in the middle line of the flattened dorsal surface, and forms a shallow longitudinal groove, is the beginning of the central nervous system, the medullary tube. Thus the primary outer germinal layer divides into two parts, the middle medullary plate (Figure 1.81 mp) and the horny–plate (ak), the beginning of the outer skin or epidermis. As the parallel borders of the concave medullary plate fold towards each other and grow underneath the horny–plate, a cylindrical tube is formed, the medullary tube (Figure 1.82 n); this quickly detaches itself altogether from the horny–plate. At each side of the medullary tube, between it and the alimentary tube (Figures 1.79 to 1.82 dh), the two parallel longitudinal folds grow out of the dorsal wall of the alimentary tube, and these form the two coelom–pouches (Figures 1.80 and 1.81 lh). This part of the entoderm, which thus represents the first structure of the middle germinal layer, is shown darker than the rest of the inner germinal layer in Figures 1.79 to 1.82. The edges of the folds meet, and thus form closed tubes (Figure 1.81 in section).

During this interesting process the outline of a third very important organ, the chorda or axial rod, is being formed between the two coelom–pouches. This first foundation of the skeleton, a solid cylindrical cartilaginous rod, is formed in the middle line of the dorsal primitive gut–wall, from the entodermal cell–streak that remains here between the two coelom–pouches (Figures 1.79 to 1.82 ch). The chorda appears at first in the shape of a flat longitudinal fold or a shallow groove (Figures 1.80 and 1.81); it does not become a solid cylindrical cord until after separation from the primitive gut (Figure 1.82). Hence we might say that the dorsal wall of the primitive gut forms three parallel longitudinal folds at this important period—one single fold and a pair of folds. The single middle fold becomes the chorda, and lies immediately below the groove of the ectoderm, which becomes the medullary tube; the pair of folds to the right and left lie at the sides between the former and the latter, and form the coelom–pouches. The part of the primitive gut that remains after the cutting off of these three dorsal primitive organs is the permanent gut; its entoderm is the gut–gland–layer or enteric layer.

(FIGURES 1.83 AND 1.84. Chordula of the amphioxus. Figure 1.83 median longitudinal section (seen from the left). Figure 1.84 transverse section. (From Hatschek.) In Figure 1.83 the coelom–pouches are omitted, in order to show the chordula more clearly. Figure 1.84 is rather diagrammatic. h horny–plate, m medullary tube, n wall of same (n apostrophe, dorsal, n double apostrophe, ventral), ch chorda, np neuroporus, ne canalis neurentericus, d gut–cavity, r gut dorsal wall, b gut ventral wall, z yelk–cells in the latter, u primitive mouth, o mouth–pit, p promesoblasts (primitive or polar cells of the mesoderm), w parietal layer, v visceral layer of the mesoderm, c coelom, f rest of the segmentation–cavity.

FIGURES 1.85 AND 1.86. Chordula of the amphibia (the ringed adder). (From Goette.) Figure 85 median longitudinal section (seen from the left), Figure 1.86 transverse section (slightly diagrammatic). Lettering as in Figures 1.83 and 1.84.

FIGURES 1.87 AND 1.88. Diagrammatic vertical section of coelomula–embryos of vertebrates. (From Hertwig.) Figure 1.87, vertical section THROUGH the primitive mouth, Figure 1.88, vertical section BEFORE the primitive mouth. u primitive mouth, ud primitive gut. d yelk, dk yelk–nuclei, dh gut–cavity, lh body–cavity, mp medullary plate, ch chorda plate, ak and ik outer and inner germinal layers, pb parietal and vb visceral mesoblast.

FIGURES 1.89 AND 1.90. Transverse section of coelomula embryos of triton. (From Hertwig.) Figure 1.89, section THROUGH the primitive mouth. Figure 1.90, section in front of the primitive mouth, u primitive mouth. dh gut–cavity, dz yelk–cells, dp yelk–stopper, ak outer and ik inner germinal layer, pb parietal and vb visceral middle layer, m medullary plate, ch chorda.)

I give the name of chordula or chorda–larva to the embryonic stage of the vertebrate organism which is represented by the amphioxus larva at this period (Figures 1.83 and 1.84, in the third period of development according to Hatschek). (Strabo and Plinius give the name of cordula or cordyla to young fish larvae.) I ascribe the utmost phylogenetic significance to it, as it is found in all the chorda–animals (tunicates as well as vertebrates) in essentially the same form. Although the accumulation of food–yelk greatly modifies the form of the chordula in the higher vertebrates, it remains the same in its main features throughout. In all cases the nerve–tube (m) lies on the dorsal side of the bilateral, worm–like body, the gut–tube (d) on the ventral side, the chorda (ch) between the two, on the long axis, and the coelom pouches (c) at each side. In every case these primitive organs develop in the same way from the germinal layers, and the same organs always arise from them in the mature chorda–animal. Hence we may conclude, according to the laws of the theory of descent, that all these chordonia or chordata (tunicates and vertebrates) descend from an ancient common ancestral form, which we may call Chordaea. We should regard this long–extinct Chordaea, if it were still in existence, as a special class of unarticulated worm (chordaria). It is especially noteworthy that neither the dorsal nerve–tube nor the ventral gut–tube, nor even the chorda that lies between them, shows any trace of articulation or segmentation; even the two coelom–sacs are not segmented at first (though in the amphioxus they quickly divide into a series of parts by transverse folding). These ontogenetic facts are of the greatest importance for the purpose of learning those ancestral forms of the vertebrates which we have to seek in the group of the unarticulated vermalia. The coelom–pouches were originally sexual glands in these ancient chordonia.

(FIGURE 1.91. A, B, C. Vertical section of the dorsal part of three triton–embryos. (From Hertwig.) In Figure A the medullary swellings (the parallel borders of the medullary plate) begin to rise; in Figure B they grow towards each other; in Figure C they join and form the medullary tube. mp medullary plate, mf medullary folds, n nerve–tube, ch chorda, lh body–cavity, mk1 and mk2 parietal and visceral mesoblasts, uv primitive–segment cavities, ak ectoderm, ik entoderm, dz yelk–cells, dh gut–cavity.)

From the evolutionary point of view the coelom–pouches are, in any case, older than the chorda; since they also develop in the same way as in the chordonia in a number of invertebrates which have no chorda (for instance, Sagitta, Figures 1.76 to 1.78). Moreover, in the amphioxus the first outline of the chorda appears later than that of the coelom–sacs. Hence we must, according to the biogenetic law, postulate a special intermediate form between the gastrula and the chordula, which we will call coelomula, an unarticulated, worm–like body with primitive gut, primitive mouth, and a double body–cavity, but no chorda. This embryonic form, the bilateral coelomula (Figure 1.81), may in turn be regarded as the ontogenetic reproduction (maintained by heredity) of an ancient ancestral form of the coelomaria, the Coelomaea (cf. Chapter 2.20).

In Sagitta and other worm–like animals the two coelom–pouches (presumably gonads or sex–glands) are separated by a complete median partition, the dorsal and ventral mesentery (Figure 1.78 dm and vm); but in the vertebrates only the upper part of this vertical partition is maintained, and forms the dorsal mesentery. This mesentery afterwards takes the form of a thin membrane, which fastens the visceral tube to the chorda (or the vertebral column). At the under side of the visceral tube the coelom–sacs blend together, their inner or median walls breaking down and disappearing. The body–cavity then forms a single simple hollow, in which the gut is quite free, or only attached to the dorsal wall by means of the mesentery.

The development of the body–cavity and the formation of the chordula in the higher vertebrates is, like that of the gastrula, chiefly modified by the pressure of the food–yelk on the embryonic structures, which forces its hinder part into a discoid expansion. These cenogenetic modifications seem to be so great that until twenty years ago these important processes were totally misunderstood. It was generally believed that the body–cavity in man and the higher vertebrates was due to the division of a simple middle layer, and that the latter arose by cleavage from one or both of the primary germinal layers. The truth was brought to light at last by the comparative embryological research of the Hertwigs. They showed in their Coelom Theory (1881) that all vertebrates are true enterocoela, and that in every case a pair of coelom–pouches are developed from the primitive gut by folding. The cenogenetic chordula–forms of the craniotes must therefore be derived from the palingenetic embryology of the amphioxus in the same way as I had previously proved for their gastrula–forms.

The chief difference between the coelomation of the acrania (amphioxus) and the other vertebrates (with skulls—craniotes) is that the two coelom–folds of the primitive gut in the former are from the first hollow vesicles, filled with fluid, but in the latter are empty pouches, the layers of which (inner and outer) close with each other. In common parlance we still call a pouch or pocket by that name, whether it is full or empty. It is different in ontogeny; in some of our embryological literature ordinary logic does not count for very much. In many of the manuals and large treatises on this science it is proved that vesicles, pouches, or sacs deserve that name only when they are inflated and filled with a clear fluid. When they are not so filled (for instance, when the primitive gut of the gastrula is filled with yelk, or when the walls of the empty coelom–pouches are pressed together), these vesicles must not be cavities any longer, but "solid structures."

The accumulation of food–yelk in the ventral wall of the primitive gut (Figures 1.85 and 1.86) is the simple cause that converts the sac–shaped coelom–pouches of the acrania into the leaf–shaped coelom–streaks of the craniotes. To convince ourselves of this we need only compare, with Hertwig, the palingenetic coelomula of the amphioxus (Figures 1.80 and 1.81) with the corresponding cenogenetic form of the amphibia (Figures 1.89 to 1.90), and construct the simple diagram that connects the two (Figures 1.87 and 1.88). If we imagine the ventral half of the primitive gut–wall in the amphioxus embryo (Figures 1.79 to 1.84) distended with food–yelk, the vesicular coelom–pouches (lh) must be pressed together by this, and forced to extend in the shape of a thin double plate between the gut–wall and body–wall (Figures 1.86 and 1.87). This expansion follows a downward and forward direction. They are not directly connected with these two walls. The real unbroken connection between the two middle layers and the primary germ–layers is found right at the back, in the region of the primitive mouth (Figure 1.87 u). At this important spot we have the source of embryonic development (blastocrene), or "zone of growth," from which the coelomation (and also the gastrulation) originally proceeds.

(FIGURE 1.92. Transverse section of the chordula–embryo of a bird (from a hen's egg at the close of the first day of incubation). (From Kolliker,) h horn–plate (ectoderm), m medullary plate, Rf dorsal folds of same, Pv medullary furrow, ch chorda, uwp median (inner) part of the middle layer (median wall of the coelom–pouches), sp lateral (outer) part of same, or lateral plates, uwh structure of the body–cavity, dd gut–gland–layer.)

Hertwig even succeeded in showing, in the coelomula–embryo of the water salamander (Triton), between the first structures of the two middle layers, the relic of the body–cavity, which is represented in the diagrammatic transitional form (Figures 1.87 and 1.88). In sections both through the primitive mouth itself (Figure 1.89) and in front of it (Figure 1.90) the two middle layers (pb and vb) diverge from each other, and disclose the two body–cavities as narrow clefts. At the primitive–mouth itself (Figure 1.90 u) we can penetrate into them from without. It is only here at the border of the primitive mouth that we can show the direct transition of the two middle layers into the two limiting layers or primary germinal layers.

The structure of the chorda also shows the same features in these coelomula–embryos of the amphibia (Figure 1.91) as in the amphioxus (Figures 1.79 to 1.82). It arises from the entodermic cell–streak, which forms the middle dorsal–line of the primitive gut, and occupies the space between the flat coelom–pouches (Figure 1.91 A). While the nervous centre is formed here in the middle line of the back and separated from the ectoderm as "medullary tube," there takes place at the same time, directly underneath, the severance of the chorda from the entoderm (Figure 1.91 A, B, C). Under the chorda is formed (out of the ventral entodermic half of the gastrula) the permanent gut or visceral cavity (enteron) (Figure 1.91 B, dh). This is done by the coalescence, under the chorda in the median line, of the two dorsal side–borders of the gut–gland–layer (ik), which were previously separated by the chorda–plate (Figure 1.91 A, ch); these now alone form the clothing of the visceral cavity (dh) (enteroderm, Figure 1.91 C). All these important modifications take place at first in the fore or head–part of the embryo, and spread backwards from there; here at the hinder end, the region of the primitive mouth, the important border of the mouth (or properistoma) remains for a long time the source of development or the zone of fresh construction, in the further building–up of the organism. One has only to compare carefully the illustrations given (Figures 1.85 to 1.91) to see that, as a fact, the cenogenetic coelomation of the amphibia can be deduced directly from the palingenetic form of the acrania (Figures 1.79 to 1.84).

(FIGURE 1.93. Transverse section of the vertebrate–embryo of a bird (from a hen's egg on the second day of incubation). (From Kolliker.) h horn–plate, mr medullary tube, ch chorda, uw primitive segments, uwh primitive–segment cavity (median relic of the coelom), sp lateral coelom–cleft, hpl skin–fibre–layer, df gut–fibre–layer, ung primitive–kidney passage, ao primitive aorta, dd gut–gland–layer.)

The same principle holds good for the amniotes, the reptiles, birds, and mammals, although in this case the processes of coelomation are more modified and more difficult to identify on account of the colossal accumulation of food–yelk and the corresponding notable flattening of the germinal disk. However, as the whole group of the amniotes has been developed at a comparatively late date from the class of the amphibia, their coelomation must also be directly traceable to that of the latter. This is really possible as a matter of fact; even the older illustrations showed an essential identity of features. Thus forty years ago Kolliker gave, in the first edition of his Human Embryology (1861), some sections of the chicken–embryo, the features of which could at once be reduced to those already described and explained in the sense of Hertwig's coelom–theory. A section through the embryo in the hatched hen's egg towards the close of the first day of incubation shows in the middle of the dorsal surface a broad ectodermic medullary groove (Figure 1.92 Rf), and underneath the middle of the chorda (ch) and at each side of it a couple of broad mesodermic layers (sp). These enclose a narrow space or cleft (uwh), which is nothing else than the structure of the body–cavity. The two layers that enclose it—the upper parietal layer (hpl) and the lower visceral layer (df)—are pressed together from without, but clearly distinguishable. This is even clearer a little later, when the medullary furrow is closed into the nerve–tube (Figure 1.93 mr).

Special importance attaches to the fact that here again the four secondary germinal layers are already sharply distinct, and easily separated from each other. There is only one very restricted area in which they are connected, and actually pass into each other; this is the region of the primitive mouth, which is contracted in the amniotes into a dorsal longitudinal cleft, the primitive groove. Its two lateral lip–borders form the primitive streak, which has long been recognised as the most important embryonic source and starting–point of further processes. Sections through this primitive streak (Figures 1.94 and 1.95) show that the two primary germinal layers grow at an early stage (in the discoid gastrula of the chick, a few hours after incubation) into the primitive streak (x), and that the two middle layers extend outward from this thickened axial plate (y) to the right and left between the former. The plates of the coelom–layers, the parietal skin–fibre–layer (m) and the visceral gut–fibre–layer (f), are seen to be still pressed close together, and only diverge later to form the body–cavity. Between the inner borders of the two flat coelom–pouches lies the chorda (Figure 1.95 x), which here again develops from the middle line of the dorsal wall of the primitive gut.

(FIGURES 1.94 AND 1.95. Transverse section of the primitive–streak (primitive mouth) of the chick. Figure 1.94 a few hours after the commencement of incubation, Figure 1.95 a little later. (From Waldeyer.) h horn–plate, n nerve–plate, m skin–fibre–layer, f gut–fibre–layer, d gut–gland–layer, y primitive streak or axial plate, in which all four germinal layers meet, x structure of the chorda, u region of the later primitive kidneys.)

Coelomation takes place in the vertebrates in just the same way as in the birds and reptiles. This was to be expected, as the characteristic gastrulation of the mammal has descended from that of the reptiles. In both cases a discoid gastrula with primitive streak arises from the segmented ovum, a two–layered germinal disk with long and small hinder primitive mouth. Here again the two primary germinal layers are only directly connected (Figure 1.96 pr) along the primitive streak (at the folding–point of the blastula), and from this spot (the border of the primitive mouth) the middle germinal layers (mk) grow out to right and left between the preceding. In the fine illustration of the coelomula of the rabbit which Van Beneden has given us (Figure 1.96) one can clearly see that each of the four secondary germinal layers consists of a single stratum of cells.

Finally, we must point out, as a fact of the utmost importance for our anthropogeny and of great general interest, that the four–layered coelomula of man has just the same construction as that of the rabbit (Figure 1.96). A vertical section that Count Spee made through the primitive mouth or streak of a very young human germinal disk (Figure 1.97) clearly shows that here again the four secondary germ–layers are inseparably connected only at the primitive streak, and that here also the two flattened coelom–pouches (mk) extend outwards to right and left from the primitive mouth between the outer and inner germinal layers. In this case, too, the middle germinal layer consists from the first of two separate strata of cells, the parietal (mp) and visceral (mv) mesoblasts.

(FIGURE 1.96. Transverse section of the primitive groove (or primitive mouth) of a rabbit. (From Van Beneden.) pr primitive mouth, ul lips of same (primitive lips), ak and ik outer and inner germinal layers, mk middle germinal layer, mp parietal layer, mv visceral layer of the mesoderm.

FIGURE 1.97. Transverse section of the primitive mouth (or groove) of a human embryo (at the coelomula stage). (From Count Spee.) pr primitive mouth, ul lips of same (primitive folds), ak and ik outer and inner germinal layers, mk middle layer, mp parietal layer, mv visceral layer of the mesoblasts.)

These concordant results of the best recent investigations (which have been confirmed by the observations of a number of scientists I have not enumerated) prove the unity of the vertebrate–stem in point of coelomation, no less than of gastrulation. In both respects the invaluable amphioxus—the sole survivor of the acrania—is found to be the original model that has preserved for us in palingenetic form by a tenacious heredity these most important embryonic processes. From this primary model of construction we can cenogenetically deduce all the embryonic forms of the other vertebrates, the craniota, by secondary modifications. My thesis of the universal formation of the gastrula by folding of the blastula has now been clearly proved for all the vertebrates; so also has been Hertwig's thesis of the origin of the middle germinal layers by the folding of a couple of coelom–pouches which appear at the border of the primitive mouth. Just as the gastraea–theory explains the origin and identity of the two primary layers, so the coelom–theory explains those of the four secondary layers. The point of origin is always the properistoma, the border of the original primitive mouth of the gastrula, at which the two primary layers pass directly into each other.

Moreover, the coelomula is important as the immediate source of the chordula, the embryonic reproduction of the ancient, typical, unarticulated, worm–like form, which has an axial chorda between the dorsal nerve–tube and the ventral gut–tube. This instructive chordula (Figures 1.83 to 1.86) provides a valuable support of our phylogeny; it indicates the important moment in our stem–history at which the stem of the chordonia (tunicates and vertebrates) parted for ever from the divergent stems of the other metazoa (articulates, echinoderms, and molluscs).

I may express here my opinion, in the form of a chordaea–theory, that the characteristic chordula–larva of the chordonia has in reality this great significance—it is the typical reproduction (preserved by heredity) of the ancient common stem–form of all the vertebrates and tunicates, the long–extinct Chordaea. We will return in Chapter 2.20 to these worm–like ancestors, which stand out as luminous points in the obscure stem–history of the invertebrate ancestors of our race.

CHAPTER 1.11. THE VERTEBRATE CHARACTER OF MAN.

We have now secured a number of firm standing–places in the labyrinthian course of our individual development by our study of the important embryonic forms which we have called the cytula, morula, blastula, gastrula, coelomula, and chordula. But we have still in front of us the difficult task of deriving the complicated frame of the human body, with all its different parts, organs, members, etc., from the simple form of the chordula. We have previously considered the origin of this four–layered embryonic form from the two–layered gastrula. The two primary germinal layers, which form the entire body of the gastrula, and the two middle layers of the coelomula that develop between them, are the four simple cell–strata, or epithelia, which alone go to the formation of the complex body of man and the higher animals. It is so difficult to understand this construction that we will first seek a companion who may help us out of many difficulties.

This helpful associate is the science of comparative anatomy. Its task is, by comparing the fully–developed bodily forms in the various groups of animals, to learn the general laws of organisation according to which the body is constructed; at the same time, it has to determine the affinities of the various groups by critical appreciation of the degrees of difference between them. Formerly, this work was conceived in a teleological sense, and it was sought to find traces of the plan of the Creator in the actual purposive organisation of animals. But comparative anatomy has gone much deeper since the establishment of the theory of descent; its philosophic aim now is to explain the variety of organic forms by adaptation, and their similarity by heredity. At the same time, it has to recognise in the shades of difference in form the degree of blood–relationship, and make an effort to construct the ancestral tree of the animal world. In this way, comparative anatomy enters into the closest relations with comparative embryology on the one hand, and with the science of classification on the other.

Now, when we ask what position man occupies among the other organisms according to the latest teaching of comparative anatomy and classification, and how man's place in the zoological system is determined by comparison of the mature bodily forms, we get a very definite and significant reply; and this reply gives us extremely important conclusions that enable us to understand the embryonic development and its evolutionary purport. Since Cuvier and Baer, since the immense progress that was effected in the early decades of the nineteenth century by these two great zoologists, the opinion has generally prevailed that the whole animal kingdom may be distributed in a small number of great divisions or types. They are called types because a certain typical or characteristic structure is constantly preserved within each of these large sections. Since we applied the theory of descent to this doctrine of types, we have learned that this common type is an outcome of heredity; all the animals of one type are blood–relatives, or members of one stem, and can be traced to a common ancestral form. Cuvier and Baer set up four of these types: the vertebrates, articulates, molluscs, and radiates. The first three of these are still retained, and may be conceived as natural phylogenetic unities, as stems or phyla in the sense of the theory of descent. It is quite otherwise with the fourth type—the radiata. These animals, little known as yet at the beginning of the nineteenth century, were made to form a sort of lumber–room, into which were cast all the lower animals that did not belong to the other three types. As we obtained a closer acquaintance with them in the course of the last sixty years, it was found that we must distinguish among them from four to eight different types. In this way the total number of animal stems or phyla has been raised to eight or twelve (cf. Chapter 2.20).

These twelve stems of the animal kingdom are, however, by no means co–ordinate and independent types, but have definite relations, partly of subordination, to each other, and a very different phylogenetic meaning. Hence they must not be arranged simply in a row one after the other, as was generally done until thirty years ago, and is still done in some manuals. We must distribute them in three subordinate principal groups of very different value, and arrange the various stems phylogenetically on the principles which I laid down in my Monograph on the Sponges, and developed in the Study of the Gastraea Theory. We have first to distinguish the unicellular animals (protozoa) from the multicellular tissue–forming (metazoa). Only the latter exhibit the important processes of segmentation and gastrulation; and they alone have a primitive gut, and form germinal layers and tissues.

The metazoa, the tissue–animals or gut–animals, then sub–divide into two main sections, according as a body–cavity is or is not developed between the primary germinal layers. We may call these the coelenteria and coelomaria, the former are often also called zoophytes or coelenterata, and the latter bilaterals. This division is the more important as the coelenteria (without coelom) have no blood and blood–vessels, nor an anus. The coelomaria (with body–cavity) have generally an anus, and blood and blood–vessels. There are four stems belonging to the coelenteria: the gastraeads ("primitive–gut animals"), sponges, cnidaria, and platodes. Of the coelomaria we can distinguish six stems: the vermalia at the bottom represent the common stem–group (derived from the platodes) of these, the other five typical stems of the coelomaria—the molluscs, echinoderms, articulates, tunicates, and vertebrates—being evolved from them.

Man is, in his whole structure, a true vertebrate, and develops from an impregnated ovum in just the same characteristic way as the other vertebrates. There can no longer be the slightest doubt about this fundamental fact, nor of the fact that all the vertebrates form a natural phylogenetic unity, a single stem. The whole of the members of this stem, from the amphioxus and the cyclostoma to the apes and man, have the same characteristic disposition, connection, and development of the central organs, and arise in the same way from the common embryonic form of the chordula. Without going into the difficult question of the origin of this stem, we must emphasise the fact that the vertebrate stem has no direct affinity whatever to five of the other ten stems; these five isolated phyla are the sponges, cnidaria, molluscs, articulates, and echinoderms. On the other hand, there are important and, to an extent, close phylogenetic relations to the other five stems—the protozoa (through the amoebae), the gastraeads (through the blastula and gastrula), the platodes and vermalia (through the coelomula), and the tunicates (through the chordula).

How we are to explain these phylogenetic relations in the present state of our knowledge, and what place is assigned to the vertebrates in the animal ancestral tree, will be considered later (Chapter 2.20). For the present our task is to make plainer the vertebrate character of man, and especially to point out the chief peculiarities of organisation by which the vertebrate stem is profoundly separated from the other eleven stems of the animal kingdom. Only after these comparative–anatomical considerations shall we be in a position to attack the difficult question of our embryology. The development of even the simplest and lowest vertebrate from the simple chordula (Figures 1.83 to 1.86) is so complicated and difficult to follow that it is necessary to understand the organic features of the fully–formed vertebrate in order to grasp the course of its embryonic evolution. But it is equally necessary to confine our attention, in this general anatomic description of the vertebrate–body, to the essential facts, and pass by all the unessential. Hence, in giving now an ideal anatomic description of the chief features of the vertebrate and its internal organisation, I omit all the subordinate points, and restrict myself to the most important characteristics.

Much, of course, will seem to the reader to be essential that is only of subordinate and secondary interest, or even not essential at all, in the light of comparative anatomy and embryology. For instance, the skull and vertebral column and the extremities are non–essential in this sense. It is true that these parts are very important PHYSIOLOGICALLY; but for the MORPHOLOGICAL conception of the vertebrate they are not essential, because they are only found in the higher, not the lower, vertebrates. The lowest vertebrates have neither skull nor vertebrae, and no extremities or limbs. Even the human embryo passes through a stage in which it has no skull or vertebrae; the trunk is quite simple, and there is yet no trace of arms and legs. At this stage of development man, like every other higher vertebrate, is essentially similar to the simplest vertebrate form, which we now find in only one living specimen. This one lowest vertebrate that merits the closest study—undoubtedly the most interesting of all the vertebrates after man—is the famous lancelet or amphioxus, to which we have already often referred. As we are going to study it more closely later on (Chapters 2.16 and 2.17), I will only make one or two passing observations on it here.

The amphioxus lives buried in the sand of the sea, is about one or two inches in length, and has, when fully developed, the shape of a very simple, longish, lancet–like leaf; hence its name of the lancelet. The narrow body is compressed on both sides, almost equally pointed at the fore and hind ends, without any trace of external appendages or articulation of the body into head, neck, breast, abdomen, etc. Its whole shape is so simple that its first discoverer thought it was a naked snail. It was not until much later—half a century ago—that the tiny creature was studied more carefully, and was found to be a true vertebrate. More recent investigations have shown that it is of the greatest importance in connection with the comparative anatomy and ontogeny of the vertebrates, and therefore with human phylogeny. The amphioxus reveals the great secret of the origin of the vertebrates from the invertebrate vermalia, and in its development and structure connects directly with certain lower tunicates, the ascidia.

When we make a number of sections of the body of the amphioxus, firstly vertical longitudinal sections through the whole body from end to end, and secondly transverse sections from right to left, we get anatomic pictures of the utmost instructiveness (cf. Figures 1.98 to 1.102). In the main they correspond to the ideal which we form, with the aid of comparative anatomy and ontogeny, of the primitive type or build of the vertebrate—the long–extinct form to which the whole stem owes its origin. As we take the phylogenetic unity of the vertebrate stem to be beyond dispute, and assume a common origin from a primitive stem–form for all the vertebrates, from amphioxus to man, we are justified in forming a definite morphological idea of this primitive vertebrate (Prospondylus or Vertebraea). We need only imagine a few slight and unessential changes in the real sections of the amphioxus in order to have this ideal anatomic figure or diagram of the primitive vertebrate form, as we see in Figures 1.98 to 1.102. The amphioxus departs so little from this primitive form that we may, in a certain sense, describe it as a modified "primitive vertebrate."* (* The ideal figure of the vertebrate as given in Figures 1.98 to 1.102 is a hypothetical scheme or diagram, that has been chiefly constructed on the lines of the amphioxus, but with a certain attention to the comparative anatomy and ontogeny of the ascidia and appendicularia on the one hand, and of the cyclostoma and selachii on the other. This diagram has no pretension whatever to be an "exact picture," but merely an attempt to reconstruct hypothetically the unknown and long extinct vertebrate stem–form, an ideal "archetype.")

The outer form of our hypothetical primitive vertebrate was at all events very simple, and probably more or less similar to that of the lancelet. The bilateral or bilateral–symmetrical body is stretched out lengthways and compressed at the sides (Figures 1.98 to 1.100), oval in section (Figures 1.101 and 1.102). There are no external articulation and no external appendages, in the shape of limbs, legs, or fins. On the other hand, the division of the body into two sections, head and trunk, was probably clearer in

Prospondylus than it is in its little–changed ancestor, the amphioxus. In both animals the fore or head–half of the body contains different organs from the trunk, and different on the dorsal from on the ventral side. As this important division is found even in the sea–squirt, the remarkable invertebrate stem–relative of the vertebrates, we may assume that it was also found in the prochordonia, the common ancestors of both stems. It is also very pronounced in the young larvae of the cyclostoma; this fact is particularly interesting, as this palingenetic larva–form is in other respects also an important connecting–link between the higher vertebrates and the acrania.

(FIGURES 1.98 TO 1.102. The ideal primitive vertebrate (prospondylus). Diagram. Figure 1.98 side–view (from the left). Figure 1.99 back–view. Figure 1.100 front view. Figure 1.101 transverse section through the head (to the left through the gill–pouches, to the right through the gill–clefts). Figure 1.102 transverse section of the trunk (to the right a pro–renal canal is affected). a aorta, af anus, au eye, b lateral furrow (primitive renal process), c coeloma (body–cavity), d small intestine, e parietal eye (epiphysis), f fin border of the skin, g auditory vesicle, gh brain, h heart, i muscular cavity (dorsal coelom–pouch), k gill–grut, ka gill–artery, kg gill–arch, ks gill–folds, l liver, ma stomach, md mouth, ms muscles, na nose (smell pit), n renal canals, u apertures of same, o outer skin, p gullet, r spinal marrow, a sexual glands (gonads), t corium, u kidney–openings (pores of the lateral furrow), v visceral vein (chief vein). x chorda, y hypophysis (urinary appendage), z gullet–groove or gill–groove (hypobranchial groove).)

The head of the acrania, or the anterior half of the body (both of the real amphioxus and the ideal prospondylus), contains the branchial (gill) gut and heart in the ventral section and the brain and sense–organs in the dorsal section. The trunk, or posterior half of the body, contains the hepatic (liver) gut and sexual–glands in the ventral part, and the spinal marrow and most of the muscles in the dorsal part.

In the longitudinal section of the ideal vertebrate (Figure 1.98) we have in the middle of the body a thin and flexible, but stiff, cylindrical rod, pointed at both ends (ch). It goes the whole length through the middle of the body, and forms, as the central skeletal axis, the original structure of the later vertebral column. This is the axial rod, or chorda dorsalis, also called chorda vertebralis, vertebral cord, axial cord, dorsal cord, notochorda, or, briefly, chorda. This solid, but flexible and elastic, axial rod consists of a cartilaginous mass of cells, and forms the inner axial skeleton or central frame of the body; it is only found in vertebrates and tunicates, not in any other animals. As the first

161

structure of the spinal column it has the same radical significance in all vertebrates, from the amphioxus to man. But it is only in the amphioxus and the cyclostoma that the axial rod retains its simplest form throughout life. In man and all the higher vertebrates it is found only in the earlier embryonic period, and is afterwards replaced by the articulated vertebral column.

The axial rod or chorda is the real solid chief axis of the vertebrate body, and at the same time corresponds to the ideal long–axis, and serves to direct us with some confidence in the orientation of the principal organs. We therefore take the vertebrate–body in its original, natural disposition, in which the long–axis lies horizontally, the dorsal side upward and the ventral side downward (Figure 1.98). When we make a vertical section through the whole length of this long axis, the body divides into two equal and symmetrical halves, right and left. In each half we have ORIGINALLY the same organs in the same disposition and connection; only their disposal in relation to the vertical plane of section, or median plane, is exactly reversed: the left half is the reflection of the right. We call the two halves antimera (opposed–parts). In the vertical plane of section that divides the two halves the sagittal ("arrow") axis, or "dorsoventral axis," goes from the back to the belly, corresponding to the sagittal seam of the skull. But when we make a horizontal longitudinal section through the chorda, the whole body divides into a dorsal and a ventral half. The line of section that passes through the body from right to left is the transverse, frontal, or lateral axis.

The two halves of the vertebrate body that are separated by this horizontal transverse axis and by the chorda have quite different characters. The dorsal half is mainly the animal part of the body, and contains the greater part of what are called the animal organs, the nervous system, muscular system, osseous system, etc.—the instruments of movement and sensation. The ventral half is essentially the vegetative half of the body, and contains the greater part of the vertebrate's vegetal organs, the visceral and vascular systems, sexual system, etc.—the instruments of nutrition and reproduction. Hence in the construction of the dorsal half it is chiefly the outer, and in the construction of the ventral half chiefly the inner, germinal layer that is engaged. Each of the two halves develops in the shape of a tube, and encloses a cavity in which another tube is found. The dorsal half contains the narrow spinal–column cavity or vertebral canal ABOVE the chorda, in which lies the tube–shaped central nervous system, the medullary tube. The ventral half contains the much more spacious visceral cavity or body–cavity UNDERNEATH the chorda, in which we find the alimentary canal and all its appendages.

The medullary tube, as the central nervous system or psychic organ of the vertebrate is called in its first stage, consists, in man and all the higher vertebrates, of two different parts: the large brain, contained in the skull, and the long spinal cord which stretches from there over the whole dorsal part of the trunk. Even in the primitive vertebrate this composition is plainly indicated. The fore half of the body, which corresponds to the head, encloses a knob–shaped vesicle, the brain (gh); this is prolonged backwards into the thin cylindrical tube of the spinal marrow (r). Hence we find here this very important psychic organ, which accomplishes sensation, will, and thought, in the vertebrates, in its simplest form. The thick wall of the nerve–tube, which runs through the long axis of the body immediately over the axial rod, encloses a narrow central canal filled with fluid (Figures 1.98 to 1.102 r). We still find the medullary tube in this very simple form for a time in the embryo of all the vertebrates, and it retains this form in the amphioxus throughout life; only in the latter case the cylindrical medullary tube barely indicates the separation of brain and spinal cord. The lancelet's medullary tube runs nearly the whole length of the body, above the chorda, in the shape of a long thin tube of almost equal diameter throughout, and there is only a slight swelling of it right at the front to represent the rudiment of a cerebral lobe. It is probable that this peculiarity of the amphioxus is connected with the partial atrophy of its head, as the ascidian larvae on the one hand and the young cyclostoma on the other clearly show a division of the vesicular brain, or head marrow, from the thinner, tubular spinal marrow.

Probably we must trace to the same phylogenetic cause the defective nature of the sense organs of the amphioxus, which we will describe later (Chapter 2.16). Prospondylus, on the other hand, probably had three pairs of sense–organs, though of a simple character, a pair of, or a single olfactory depression, right in front (Figures 1.98 and 1.99, na), a pair of eyes (au) in the lateral walls of the brain, and a pair of simple auscultory vesicles (g) behind. There was also, perhaps, a single parietal or "pineal" eye at the top of the skull (epiphysis, e).

In the vertical median plane (or middle plane, dividing the bilateral body into right and left halves) we have in the acrania, underneath the chorda, the mesentery and visceral tube, and above it the medullary tube; and above the latter a membranous partition of the two halves of the body. With this partition is connected the mass of connective tissue which acts as a sheath both for the medullary tube and the underlying chorda, and is, therefore, called the chord–sheath (perichorda); it originates from the dorsal and median part of the coelom–pouches, which we shall call the skeleton plate or "sclerotom" in the craniote embryo. In the latter the chief part of the skeleton—the vertebral column and skull—develops from this chord–sheath; in the acrania it retains its simple form as a soft connective matter, from which are formed the membranous partitions between the various muscular plates or myotomes (Figures 1.98 and 1.99 ms).

To the right and left of the cord–sheath, at each side of the medullary tube and the underlying axial rod, we find in all the vertebrates the large masses of muscle that constitute the musculature of the trunk and effect its movements. Although these are very elaborately differentiated and connected in the developed vertebrate (corresponding to the various parts of the bony skeleton), in our ideal primitive vertebrate we can distinguish only two pairs of these principal muscles, which run the whole length of the body parallel to the chorda. These are the upper (dorsal) and lower (ventral) lateral muscles of the trunk. The upper (dorsal) muscles, or the original dorsal muscles (Figure 1.102 ms), form the thick mass of flesh on the back. The lower (ventral) muscles, or the original muscles of the belly, form the fleshy wall of the abdomen. Both sets are segmented, and consist of a double row of muscular plates (Figures 1.98 and 1.99 ms); the number of these myotomes determines the number of joints in the trunk, or metamera. The myotomes are also developed from the thick wall of the coelom–pouches (Figure 1.102 i).

Outside this muscular tube we have the external envelope of the vertebrate body, which is known as the corium or cutis. This strong and thick envelope consists, in its deeper strata, chiefly of fat and loose connective tissue, and in its upper layers of cutaneous muscles and firmer connective tissue. It covers the whole surface of the fleshy body, and is of considerable thickness in all the craniota. But in the acrania the corium is merely a thin plate of connective tissue, an insignificant "corium–plate" (lamella corii, Figures 1.98 to 1.102 t).

Immediately above the corium is the outer skin (epidermis, o), the general covering of the whole outer surface. In the higher vertebrates the hairs, nails, feathers, claws, scales, etc., grow out of this epidermis. It consists, with all its appendages and products, of simple cells, and has no blood–vessels. Its cells are connected with the terminations of the sensory nerves. Originally, the outer skin is a perfectly simple covering of the outer surface of the body, composed only of homogeneous cells—a permanent horn–plate. In this simplest form, as a one–layered epithelium, we find it, at first, in all the vertebrates, and throughout life in the acrania. It afterwards grows thicker in the higher vertebrates, and divides into two strata—an outer, firmer corneous (horn) layer and an inner, softer mucus—layer; also a number of external and internal appendages grow out of it: outwardly, the hairs, nails, claws, etc., and inwardly, the sweat—glands, fat—glands, etc.

It is probable that in our primitive vertebrate the skin was raised in the middle line of the body in the shape of a vertical fin border (f). A similar fringe, going round the greater part of the body, is found to—day in the amphioxus and the cyclostoma; we also find one in the tail of fish—larvae and tadpoles.

Now that we have considered the external parts of the vertebrate and the animal organs, which mainly lie in the dorsal half, above the chorda, we turn to the vegetal organs, which lie for the most part in the ventral half, below the axial rod. Here we find a large body—cavity or visceral cavity in all the craniota. The spacious cavity that encloses the greater part of the viscera corresponds to only a part of the original coeloma, which we considered in Chapter 1.10; hence it nay be called the metacoeloma. As a rule, it is still briefly called the coeloma; formerly it was known in anatomy as the pleuroperitoneal cavity. In man and the other mammals (but only in these) this coeloma divides, when fully developed, into two different cavities, which are separated by a transverse partition—the muscular diaphragm. The fore or pectoral cavity (pleura—cavity) contains the oesophagus (gullet), heart, and lungs; the hind or peritoneal or abdominal cavity contains the stomach, small and large intestines, liver, pancreas, kidneys, etc. But in the vertebrate embryo, before the diaphragm is developed, the two cavities form a single continuous body—cavity, and we find it thus in all the lower vertebrates throughout life. This body—cavity is clothed with a delicate layer of cells, the coelom—epithelium. In the acrania the coelom is segmented both dorsally and ventrally, as their muscular pouches and primitive genital organs plainly show (Figure 1.102).

The chief of the viscera in the body—cavity is the alimentary canal, the organ that represents the whole body in the gastrula. In all the vertebrates it is a long tube, enclosed in the body—cavity and more or less differentiated in length, and has two apertures—a mouth for taking in food (Figures 1.98 and 1.100 md) and an anus for the ejection of unusable matter or excrements (af). With the alimentary canal a number of glands are connected which are of great importance for the vertebrate body, and which all grow out of the canal. Glands of this kind are the salivary glands, the lungs, the liver, and many smaller glands. Nearly all these glands are wanting in the acrania; probably there were merely a couple of simple hepatic tubes (Figures 1.98 and 1.100 l) in the vertebrate stem—form. The wall of the alimentary canal and all its appendages consists of two different layers; the inner, cellular clothing is the gut–gland–layer, and the outer, fibrous envelope consists of the gut–fibre–layer; it is mainly composed of muscular fibres which accomplish the digestive movements of the canal, and of connective–tissue fibres that form a firm envelope. We have a continuation of it in the mesentery, a thin, bandage–like layer, by means of which the alimentary canal is fastened to the ventral side of the chorda, originally the dorsal partition of the two coelom–pouches. The alimentary canal is variously modified in the vertebrates both as a whole and in its several sections, though the original structure is always the same, and is very simple. As a rule, it is longer (often several times longer) than the body, and therefore folded and winding within the body–cavity, especially at the lower end. In man and the higher vertebrates it is divided into several sections, often separated by valves—the mouth, pharynx, oesophagus, stomach, small and large intestine, and rectum. All these parts develop from a very simple structure, which originally (throughout life in the amphioxus) runs from end to end under the chorda in the shape of a straight cylindrical canal.

As the alimentary canal may be regarded morphologically as the oldest and most important organ in the body, it is interesting to understand its essential features in the vertebrate more fully, and distinguish them from unessential features. In this connection we must particularly note that the alimentary canal of every vertebrate shows a very characteristic division into two sections—a fore and a hind chamber. The fore chamber is the head–gut or branchial gut (Figures 1.98 to 1.100 p, k), and is chiefly occupied with respiration. The hind section is the trunk–gut or hepatic gut, which accomplishes digestion (ma, d). In all vertebrates there are formed, at an early stage, to the right and left in the fore–part of the head–gut, certain special clefts that have an intimate connection with the original respiratory apparatus of the vertebrate—the branchial (gill) clefts (ks). All the lower vertebrates, the lancelets, lampreys, and fishes, are constantly taking in water at the mouth, and letting it out again by the lateral clefts of the gullet. This water serves for breathing. The oxygen contained in it is inspired by the blood–canals, which spread out on the parts between the gill–clefts, the gill–arches (kg). These very characteristic branchial clefts and arches are found in the embryo of man and all the higher vertebrates at an early stage of development, just as we find them throughout life in the lower vertebrates. However, these clefts and arches never act as respiratory organs in the mammals, birds, and reptiles, but gradually develop into quite different parts. Still, the fact that they are found at first in the same form as in the fishes is one of the most interesting proofs of the descent of these three higher classes from the fishes.

Not less interesting and important is an organ that develops from the ventral wall in all vertebrates—the gill–groove or hypobranchial groove. In the acrania and the ascidiae it consists throughout life of a glandular ciliated groove, which runs down from the mouth in the ventral middle line of the gill–gut, and takes small particles of food to the stomach (Figure 1.101 z). But in the craniota the thyroid gland (thyreoidea) is developed from it, the gland that lies in front of the larynx, and which, when pathologically enlarged, forms goitre (struma).

From the head–gut we get not only the gills, the organs of water–breathing in the lower vertebrates, but also the lungs, the organs of atmospheric breathing in the five higher classes. In these cases a vesicular fold appears in the gullet of the embryo at an early stage, and gradually takes the shape of two spacious sacs, which are afterwards filled with air. These sacs are the two air–breathing lungs, which take the place of the water–breathing gills. But the vesicular invagination, from which the lungs arise, is merely the familiar air–filled vesicle, which we call the floating–bladder of the fish, and which alters its specific weight, acting as hydrostatic organ or floating apparatus. This structure is not found in the lowest vertebrate classes—the acrania and cyclostoma. We shall see more of it in Volume 2.

The second chief section of the vertebrate–gut, the trunk or liver–gut, which accomplishes digestion, is of very simple construction in the acrania. It consists of two different chambers. The first chamber, immediately behind the gill–gut, is the expanded stomach (ma); the second, narrower and longer chamber, is the straight small intestine (d): it issues behind on the ventral side by the anus (af). Near the limit of the two chambers in the visceral cavity we find the liver, in the shape of a simple tube or blind sac (l); in the amphioxus it is single; in the prospondylus it was probably double (Figures 1.98 and 1.100 l).

Closely related morphologically and physiologically to the alimentary canal is the vascular system of the vertebrate, the chief sections of which develop from the fibrous gut–layer. It consists of two different but directly connected parts, the system of blood–vessels and that of lymph–vessels. In the passages of the one we find red blood, and in the other colourless lymph. To the lymphatic system belong, first of all, the lymphatic canals proper or absorbent veins, which are distributed among all the organs, and absorb the used–up juices from the tissues, and conduct them into the venous blood; but besides these there are the chyle–vessels, which absorb the white chyle, the milky fluid prepared by the alimentary canal from the food, and conduct this also to the blood.

The blood–vessel system of the vertebrate has a very elaborate construction, but seems to have had a very simple form in the primitive vertebrate, as we find it to–day permanently in the annelids (for instance, earth–worms) and the amphioxus. We accordingly distinguish first of all as essential, original parts of it two large single blood–canals, which lie in the fibrous wall of the gut, and run along the alimentary canal in the median plane of the body, one above and the other underneath the canal. These principal canals give out numerous branches to all parts of the body, and pass into each other by arches before and behind; we will call them the primitive artery and the primitive vein. The first corresponds to the dorsal vessel, the second to the ventral vessel, of the worms. The primitive or principal artery, usually called the aorta (Figure 1.98 a), lies above the gut in the middle line of its dorsal side, and conducts oxidised or arterial blood from the gills to the body. The primitive or principal vein (Figure 1.100 v) lies below the gut, in the middle line of its ventral side, and is therefore also called the vena subintestinalis; it conducts carbonised or venous blood back from the body to the gills. At the branchial section of the gut in front the two canals are connected by a number of branches, which rise in arches between the gill–clefts. These "branchial vascular arches" (kg) run along the gill–arches, and have a direct share in the work of respiration. The anterior continuation of the principal vein which runs on the ventral wall of the gill–gut, and gives off these vascular arches upwards, is the branchial artery (ka). At the border of the two sections of the ventral vessel it enlarges into a contractile spindle–shaped tube (Figures 1.98 and 1.100 h). This is the first outline of the heart, which afterwards becomes a four–chambered pump in the higher vertebrates and man. There is no heart in the amphioxus, probably owing to degeneration. In prospondylus the ventral gill–heart probably had the simple form in which we still find it in the ascidia and the embryos of the craniota (Figures 1.98 and 1.100 h).

The kidneys, which act as organs of excretion or urinary organs in all vertebrates, have a very different and elaborate construction in the various sections of this stem; we will consider them further in Chapter 2.29. Here I need only mention that in our hypothetical primitive vertebrate they probably had the same form as in the actual amphioxus—the primitive kidneys (protonephra). These are originally made up of a double row of little canals, which directly convey the used–up juices or the urine out of the body–cavity (Figure 1.102 n). The inner aperture of these pronephridial canals opens with a ciliated funnel into the body–cavity; the external aperture opens in lateral grooves of the epidermis, a couple of longitudinal grooves in the lateral surface of the outer skin (Figure 1.102 b). The pronephridial duct is formed by the closing of this groove to the right and left at the sides. In all the craniota it develops at an early stage in the horny plate; in the amphioxus it seems to be converted into a wide cavity, the atrium, or peribranchial space.

Next to the kidneys we have the sexual organs of the vertebrate. In most of the members of this stem the two are united in a single urogenital system; it is only in a few groups that the urinary and sexual organs are separated (in the amphioxus, the cyclostoma, and some sections of the fish–class). In man and all the higher vertebrates the sexual apparatus is made up of various parts, which we will consider in Chapter 2.29. But in the two lowest classes of our stem, the acrania and cyclostoma, they consist merely of simple sexual glands or gonads, the ovaries of the female sex and the testicles (spermaria) of the male; the former provide the ova, the latter the sperm. In the craniota we always find only one pair of gonads; in the amphioxus several pairs, arranged in succession. They must have had the same form in our hypothetical prospondylus (Figures 1.98 and 1.100 s). These segmental pairs of gonads are the original ventral halves of the coelom–pouches.

The organs which we have now enumerated in this general survey, and of which we have noted the characteristic disposition, are those parts of the organism that are found in all vertebrates without exception in the same relation to each other, however much they may be modified. We have chiefly had in view the transverse section of the body (Figures 1.101 and 1.102), because in this we see most clearly the distinctive arrangement of them. But to complete our picture we must also consider the segmentation or metamera–formation of them, which has yet been hardly noticed, and which is seen best in the longitudinal section. In man and all the more advanced vertebrates the body is made up of a series or chain of similar members, which succeed each other in the long axis of the body—the segments or metamera of the organism. In man these homogeneous parts number thirty–three in the trunk, but they run to several hundred in many of the vertebrates (such as serpents or eels). As this internal articulation or metamerism is mainly found in the vertebral column and the surrounding muscles, the sections or metamera were formerly called pro–vertebrae. As a fact, the articulation is by no means chiefly determined and caused by the skeleton, but by the muscular system and the segmental arrangement of the kidneys and gonads. However, the composition from these pro–vertebrae or internal metamera is usually, and rightly, put forward as a prominent character of the vertebrate, and the manifold division or differentiation of them is of great importance in the various groups of the vertebrates. But as far as our present task—the derivation of the simple body of the primitive vertebrate from the chordula—is concerned, the articulate parts or metamera are of secondary interest, and we need not go into them just now.

(FIGURE 1.103 A, B, C, D. Instances of redundant mammary glands and nipples (hypermastism). A a pair of small redundant breasts (with two nipples on the left) above the large normal ones; from a 45–year–old Berlin woman, who had had children 17 times (twins twice). (From Hansemann.) B the highest number: ten nipples (all giving milk), three pairs above, one pair below, the large normal breasts; from a 22–year–old servant at Warschau. (From Neugebaur.) C three pairs of nipples: two pairs on the normal glands and one pair above; from a 19–year–old Japanese girl. D four pairs of nipples: one pair above the normal and two pairs of small accessory nipples underneath; from a 22–year–old Bavarian soldier. (From Wiedersheim.))

The characteristic composition of the vertebrate body develops from the embryonic structure in the same way in man as in all the other vertebrates. As all competent experts now admit the monophyletic origin of the vertebrates on the strength of this significant agreement, and this "common descent of all the vertebrates from one original stem–form" is admitted as an historical fact, we have found the answer to "the question of questions." We may, moreover, point out that this answer is just as certain and precise in the case of the origin of man from the mammals. This advanced vertebrate class is also monophyletic, or has evolved from one common stem–group of lower vertebrates (reptiles, and, earlier still, amphibia). This follows from the fact that the mammals are clearly distinguished from the other classes of the stem, not merely in one striking particular, but in a whole group of distinctive characters.

It is only in the mammals that we find the skin covered with hair, the breast–cavity separated from the abdominal cavity by a complete diaphragm, and the larynx provided with an epiglottis. The mammals alone have three small auscultory bones in the tympanic cavity—a feature that is connected with the characteristic modification of their maxillary joint. Their red blood–cells have no nucleus, whereas this is retained in all other vertebrates. Finally, it is only in the mammals that we find the remarkable function of the breast structure which has given its name to the whole class—the feeding of the young by the mother's milk. The mammary glands which serve this purpose are interesting in so many ways that we may devote a few lines to them here.

As is well known, the lower mammals, especially those which beget a number of young at a time, have several mammary glands at the breast. Hedgehogs and sows have five pairs, mice four or five pairs, dogs and squirrels four pairs, cats and bears three pairs, most of the ruminants and many of the rodents two pairs, each provided with a teat or nipple (mastos). In the various genera of the half—apes (lemurs) the number varies a good deal. On the other hand, the bats and apes, which only beget one young at a time as a rule, have only one pair of mammary glands, and these are found at the breast, as in man.

These variations in the number or structure of the mammary apparatus (mammarium) have become doubly interesting in the light of recent research in comparative anatomy. It has been shown that in man and the apes we often find redundant mammary glands (hyper—mastism) and corresponding teats (hyper—thelism) in both sexes. Figure 1.103 shows four cases of this kind—A, B, and C of three women, and D of a man. They prove that all the above—mentioned numbers may be found occasionally in man. Figure 1.103 A shows the breast of a Berlin woman who had had children seventeen times, and who has a pair of small accessory breasts (with two nipples on the left one) above the two normal breasts; this is a common occurrence, and the small soft pad above the breast is not infrequently represented in ancient statues of Venus. In Figure 1.103 C we have the same phenomenon in a Japanese girl of nineteen, who has two nipples on each breast besides (three pairs altogether). Figure 1.103 D is a man of twenty—two with four pairs of nipples (as in the dog), a small pair above and two small pairs beneath the large normal teats. The maximum number of five pairs (as in the sow and hedgehog) was found in a Polish servant of twenty—two who had had several children; milk was given by each nipple; there were three pairs of redundant nipples above and one pair underneath the normal and very large breasts (Figure 1.103 B).

A number of recent investigations (especially among recruits) have shown that these things are not uncommon in the male as well as the female sex. They can only be explained by evolution, which attributes them to atavism and latent heredity. The earlier ancestors of all the primates (including man) were lower placentals, which had, like the hedgehog (one of the oldest forms of the living placentals), several mammary glands (five or more pairs) in the abdominal skin. In the apes and man only a couple of them are normally developed, but from time to time we get a development of the atrophied structures. Special notice should be taken of the arrangement of these accessory mammae; they form, as is clearly seen in Figure 1.103 B and D, two long rows, which diverge forward (towards the arm—pit), and converge behind in the middle line (towards the loins). The milk–glands of the polymastic lower placentals are arranged in similar lines.

The phylogenetic explanation of polymastism, as given in comparative anatomy, has lately found considerable support in ontogeny. Hans Strahl, E. Schmitt, and others, have found that there are always in the human embryo at the sixth week (when it is three–fifths of an inch long) the microscopic traces of five pairs of mammary glands, and that they are arranged at regular distances in two lateral and divergent lines, which correspond to the mammary lines. Only one pair of them—the central pair—are normally developed, the others atrophying. Hence there is for a time in the human embryo a normal hyperthelism, and this can only be explained by the descent of man from lower primates (lemurs) with several pairs.

But the milk–gland of the mammal has a great morphological interest from another point of view. This organ for feeding the young in man and the higher mammals is, as is known, found in both sexes. However, it is usually active only in the female sex, and yields the valuable "mother's milk"; in the male sex it is small and inactive, a real rudimentary organ of no physiological interest. Nevertheless, in certain cases we find the breast as fully developed in man as in woman, and it may give milk for feeding the young.

(FIGURE 1.104. A Greek gynecomast.)

We have a striking instance of this gynecomastism (large milk–giving breasts in a male) in Figure 1.104. I owe the photograph (taken from life) to the kindness of Dr. Ornstein, of Athens, a German physician, who has rendered service by a number of anthropological observations, (for instance, in several cases of tailed men). The gynecomast in question is a Greek recruit in his twentieth year, who has both normally developed male organs and very pronounced female breasts. It is noteworthy that the other features of his structure are in accord with the softer forms of the female sex. It reminds us of the marble statues of hermaphrodites which the ancient Greek and Roman sculptors often produced. But the man would only be a real hermaphrodite if he had ovaries internally besides the (externally visible) testicles.

I observed a very similar case during my stay in Ceylon (at Belligemma) in 1881. A young Cinghalese in his twenty–fifth year was brought to me as a curious hermaphrodite, half–man and half–woman. His large breasts gave plenty of milk; he was employed as "male nurse" to suckle a new–born infant whose mother had died at birth. The outline of his body was softer and more feminine than in the Greek shown in Figure 1.104. As the Cinghalese are small of stature and of graceful build, and as the men often resemble the women in clothing (upper part of the body naked, female dress on the lower part) and the dressing of the hair (with a comb), I first took the beardless youth to be a woman. The illusion was greater, as in this remarkable case gynecomastism was associated with cryptorchism—that is to say, the testicles had kept to their original place in the visceral cavity, and had not travelled in the normal way down into the scrotum. (Cf. Chapter 2.29.) Hence the latter was very small, soft, and empty. Moreover, one could feel nothing of the testicles in the inguinal canal. On the other hand, the male organ was very small, but normally developed. It was clear that this apparent hermaphrodite also was a real male.

Another case of practical gynecomastism has been described by Alexander von Humboldt. In a South American forest he found a solitary settler whose wife had died in child–birth. The man had laid the new–born child on his own breast in despair; and the continuous stimulus of the child's sucking movements had revived the activity of the mammary glands. It is possible that nervous suggestion had some share in it. Similar cases have been often observed in recent years, even among other male mammals (such as sheep and goats).

The great scientific interest of these facts is in their bearing on the question of heredity. The stem–history of the mammarium rests partly on its embryology (Chapter 2.24.) and partly on the facts of comparative anatomy and physiology. As in the lower and higher mammals (the monotremes, and most of the marsupials) the whole lactiferous apparatus is only found in the female; and as there are traces of it in the male only in a few younger marsupials, there can be no doubt that these important organs were originally found only in the female mammal, and that they were acquired by these through a special adaptation to habits of life.

Later, these female organs were communicated to both sexes by heredity; and they have been maintained in all persons of either sex, although they are not physiologically active in the males. This normal permanence of the female lactiferous organs in BOTH sexes of the higher mammals and man is independent of any selection, and is a fine instance of the much–disputed "inheritance of acquired characters."

CHAPTER 1.12. EMBRYONIC SHIELD AND GERMINATIVE AREA.

The three higher classes of vertebrates which we call the amniotes—the mammals, birds, and reptiles—are notably distinguished by a number of peculiarities of their development from the five lower classes of the stem—the animals without an amnion (the anamnia). All the amniotes have a distinctive embryonic membrane known as the amnion (or "water—membrane"), and a special embryonic appendage—the allantois. They have, further, a large yelk—sac, which is filled with food—yelk in the reptiles and birds, and with a corresponding clear fluid in the mammals. In consequence of these later—acquired structures, the original features of the development of the amniotes are so much altered that it is very difficult to reduce them to the palingenetic embryonic processes of the lower amnion—less vertebrates. The gastraea theory shows us how to do this, by representing the embryology of the lowest vertebrate, the skull—less amphioxus, as the original form, and deducing from it, through a series of gradual modifications, the gastrulation and coelomation of the craniota.

It was somewhat fatal to the true conception of the chief embryonic processes of the vertebrate that all the older embryologists, from Malpighi (1687) and Wolff (1750) to Baer (1828) and Remak (1850), always started from the investigation of the hen's egg, and transferred to man and the other vertebrates the impressions they gathered from this. This classical object of embryological research is, as we have seen, a source of dangerous errors. The large round food—yelk of the bird's egg causes, in the first place, a flat discoid expansion of the small gastrula, and then so distinctive a development of this thin round embryonic disk that the controversy as to its significance occupies a large part of embryological literature.

(FIGURE 1.105. Severance of the discoid mammal embryo from the yelk—sac, in transverse section (diagrammatic). A The germinal disk (h, hf) lies flat on one side of the branchial—gut vesicle (kb). B In the middle of the germinal disk we find the medullary groove (mr), and underneath it the chorda (ch). C The gut—fibre—layer (df) has been enclosed by the gut—gland—layer (dd). D The skin—fibre—layer (hf) and gut—fibre—layer (df) divide at the periphery; the gut (d) begins to separate from the yelk—sac or umbilical vesicle (nb). E The medullary tube (mr) is closed; the body—cavity (c) begins to form. F The provertebrae (w) begin to grow round the medullary tube (mr) and the chorda (ch):

the gut (d) is cut off from the umbilical vesicle (nb). H The vertebrae (w) have grown round the medullary tube (mr) and chorda; the body–cavity is closed, and the umbilical vesicle has disappeared. The amnion and serous membrane are omitted. The letters have the same meaning throughout: h horn–plate, mr medullary tube, hf skin–fibre–layer, w provertebrae, ch chorda, c body–cavity or coeloma, df gut–fibre–layer, dd gut–gland–layer, d gut–cavity, nb umbilical vesicle.)

One of the most unfortunate errors that this led to was the idea of an original antithesis of germ and yelk. The latter was regarded as a foreign body, extrinsic to the real germ, whereas it is properly a part of it, an embryonic organ of nutrition. Many authors said there was no trace of the embryo until a later stage, and outside the yelk; sometimes the two–layered embryonic disk itself, at other times only the central portion of it (as distinguished from the germinative area, which we will describe presently), was taken to be the first outline of the embryo. In the light of the gastraea theory it is hardly necessary to dwell on the defects of this earlier view and the erroneous conclusions drawn from it. In reality, the first segmentation–cell, and even the stem–cell itself and all that issues therefrom, belong to the embryo. As the large original yelk–mass in the undivided egg of the bird only represents an inclosure in the greatly enlarged ovum, so the later contents of its embryonic yelk–sac (whether yet segmented or not) are only a part of the entoderm which forms the primitive gut. This is clearly shown by the ova of the amphibia and cyclostoma, which explain the transition from the yelk–less ova of the amphioxus to the large yelk–filled ova of the reptiles and birds.

It is precisely in the study of these difficult features that we see the incalculable value of phylogenetic considerations in explaining complex ontogenetic facts, and the need of separating cenogenetic phenomena from palingenetic. This is particularly clear as regards the comparative embryology of the vertebrates, because here the phylogenetic unity of the stem has been already established by the well–known facts of paleontology and comparative anatomy. If this unity of the stem, on the basis of the amphioxus, were always borne in mind, we should not have these errors constantly recurring.

In many cases the cenogenetic relation of the embryo to the food–yelk has until now given rise to a quite wrong idea of the first and most important embryonic processes in the higher vertebrates, and has occasioned a number of false theories in connection with them. Until thirty years ago the embryology of the higher vertebrates always started from the position that the first structure of the embryo is a flat, leaf–shaped disk; it was for this

175

reason that the cell–layers that compose this germinal disk (also called germinative area) are called "germinal layers." This flat germinal disk, which is round at first and then oval, and which is often described as the tread or cicatricula in the laid hen's egg, is found at a certain part of the surface of the large globular food–yelk. I am convinced that it is nothing else than the discoid, flattened gastrula of the birds. At the beginning of germination the flat embryonic disk curves outwards, and separates on the inner side from the underlying large yelk–ball. In this way the flat layers are converted into tubes, their edges folding and joining together (Figure 1.105). As the embryo grows at the expense of the food–yelk, the latter becomes smaller and smaller; it is completely surrounded by the germinal layers. Later still, the remainder of the food–yelk only forms a small round sac, the yelk–sac or umbilical vesicle (Figure 1.105 nb). This is enclosed by the visceral layer, is connected by a thin stalk, the yelk–duct, with the central part of the gut–tube, and is finally, in most of the vertebrates, entirely absorbed by this (H). The point at which this takes place, and where the gut finally closes, is the visceral navel. In the mammals, in which the remainder of the yelk–sac remains without and atrophies, the yelk–duct at length penetrates the outer ventral wall. At birth the umbilical cord proceeds from here, and the point of closure remains throughout life in the skin as the navel.

As the older embryology of the higher vertebrates was mainly based on the chick, and regarded the antithesis of embryo (or formative–yelk) and food–yelk (or yelk–sac) as original, it had also to look upon the flat leaf–shaped structure of the germinal disk as the primitive embryonic form, and emphasise the fact that hollow grooves were formed of these flat layers by folding, and closed tubes by the joining together of their edges.

This idea, which dominated the whole treatment of the embryology of the higher vertebrates until thirty years ago, was totally false. The gastraea theory, which has its chief application here, teaches us that it is the very reverse of the truth. The cup–shaped gastrula, in the body–wall of which the two primary germinal layers appear from the first as closed tubes, is the original embryonic form of all the vertebrates, and all the multicellular invertebrates; and the flat germinal disk with its superficially expanded germinal layers is a later, secondary form, due to the cenogenetic formation of the large food–yelk and the gradual spread of the germ–layers over its surface. Hence the actual folding of the germinal layers and their conversion into tubes is not an original and primary, but a much later and tertiary, evolutionary process. In the phylogeny of the vertebrate embryonic process we may distinguish the following three stages:—

A. First Stage: Primary (palingenetic) embryonic process.

The germinal layers form from the first closed tubes, the one–layered blastula being converted into the two–layered gastrula by invagination. No food–yelk. (Amphioxus.)

B. Second Stage: Secondary (cenogenetic) embryonic process.

The germinal layers spread out leaf–wise, food–yelk gathering in the ventral entoderm, and a large yelk–sac being formed from the middle of the gut–tube. (Amphibia.)

C. Third Stage: Tertiary (cenogenetic) embryonic process.

The germinal layers form a flat germinal disk, the borders of which join together and form closed tubes, separating from the central yelk–sac. (Amniotes.)

As this theory, a logical conclusion from the gastraea theory, has been fully substantiated by the comparative study of gastrulation in the last few decades, we must exactly reverse the hitherto prevalent mode of treatment. The yelk–sac is not to be treated, as was done formerly, as if it were originally antithetic to the embryo, but as an essential part of it, a part of its visceral tube. The primitive gut of the gastrula has, on this view, been divided into two parts in the higher animals as a result of the cenogenetic formation of the food–yelk—the permanent gut (metagaster), or permanent alimentary canal, and the yelk–sac (lecithoma), or umbilical vesicle. This is very clearly shown by the comparative ontogeny of the fishes and amphibia. In these cases the whole yelk undergoes cleavage at first, and forms a yelk–gland, composed of yelk–cells, in the ventral wall of the primitive gut. But it afterwards becomes so large that a part of the yelk does not divide, and is used up in the yelk–sac that is cut off outside.

(FIGURE 1.106. The visceral embryonic vesicle (blastocystis or gastrocystis) of a rabbit (the "blastula" or vesicula blastodermica of other writers), a outer envelope (ovolemma), b skin–layer or ectoderm, forming the entire wall of the yelk–vesicle, c groups of dark cells, representing the visceral layer or entoderm.

FIGURE 1.107. The same in section. Letters as above. d cavity of the vesicle. (From Bischoff.))

When we make a comparative study of the embryology of the amphioxus, the frog, the chick, and the rabbit, there cannot, in my opinion, be any further doubt as to the truth of this position, which I have held for thirty years. Hence in the light of the gastraea theory we must regard the features of the amphioxus as the only and real primitive structure among all the vertebrates, departing very little from the palingenetic embryonic form. In the cyclostoma and the frog these features are, on the whole, not much altered cenogenetically, but they are very much so in the chick, and most of all in the rabbit. In the bell–gastrula of the amphioxus and in the hooded gastrula of the lamprey and the frog the germinal layers are found to be closed tubes or vesicles from the first. On the other hand, the chick–embryo (in the new laid, but not yet hatched, egg) is a flat circular disk, and it was not easy to recognise this as a real gastrula. Rauber and Goette have, however, achieved this. As the discoid gastrula grows round the large globular yelk, and the permanent gut then separates from the outlying yelk–sac, we find all the processes which we have shown (diagrammatically) in Figure 1.108—processes that were hitherto regarded as principal acts, whereas they are merely secondary.

The oldest, oviparous mammals, the monotremes, behave in the same way as the reptiles and birds. But the corresponding embryonic processes in the viviparous mammals, the marsupials and placentals, are very elaborate and distinctive. They were formerly quite misinterpreted; it was not until the publication of the studies of Edward van Beneden (1875) and the later research of Selenka, Kuppfer, Rabl, and others, that light was thrown on them, and we were in a position to bring them into line with the principles of the gastraea theory and trace them to the embryonic forms of the lower vertebrates. Although there is no independent food–yelk, apart from the formative yelk, in the mammal ovum, and although its segmentation is total on that account, nevertheless a large yelk–sac is formed in their embryos, and the "embryo proper" spreads leaf–wise over its surface, as in the reptiles and birds, which have a large food–yelk and partial segmentation. In the mammals, as well as in the latter, the flat, leaf–shaped germinal disk separates from the yelk–sac, and its edges join together and form tubes.

How can we explain this curious anomaly? Only as a result of very characteristic and peculiar cenogenetic modifications of the embryonic process, the real causes of which must be sought in the change in the rearing of the young on the part of the viviparous mammals. These are clearly connected with the fact that the ancestors of the viviparous mammals were oviparous amniotes like the present monotremes, and only gradually became viviparous. This can no longer be questioned now that it has been shown (1884) that the monotremes, the lowest and oldest of the mammals, still lay eggs, and that these develop like the ova of the reptiles and birds. Their nearest descendants, the marsupials, formed the habit of retaining the eggs, and developing them in the oviduct; the latter was thus converted into a womb (uterus). A nutritive fluid that was secreted from its wall, and passed through the wall of the blastula, now served to feed the embryo, and took the place of the food–yelk. In this way the original food–yelk of the monotremes gradually atrophied, and at last disappeared so completely that the partial ovum–segmentation of their descendants, the rest of the mammals, once more became total. From the discogastrula of the former was evolved the distinctive epigastrula of the latter.

It is only by this phylogenetic explanation that we can understand the formation and development of the peculiar, and hitherto totally misunderstood, blastula of the mammal. The vesicular condition of the mammal embryo was discovered 200 years ago (1677) by Regner de Graaf. He found in the uterus of a rabbit four days after impregnation small, round, loose, transparent vesicles, with a double envelope. However, Graaf's discovery passed without recognition. It was not until 1827 that these vesicles were rediscovered by Baer, and then more closely studied in 1842 by Bischoff in the rabbit (Figures 1.106 and 1.107). They are found in the womb of the rabbit, the dog, and other small mammals, a few days after copulation. The mature ova of the mammal, when they have left the ovary, are fertilised either here or in the oviduct immediately afterwards by the invading sperm–cells.* (* In man and the other mammals the fertilisation of the ova probably takes place, as a rule, in the oviduct; here the ova, which issue from the female ovary in the shape of the Graafian follicle, and enter the inner aperture of the oviduct, encounter the mobile sperm–cells of the male seed, which pass into the uterus at copulation, and from this into the external aperture of the oviduct. Impregnation rarely takes place in the ovary or in the womb.) (As to the womb and oviduct see Chapter 2.29.) The cleavage and formation of the gastrula take place in the oviduct. Either here in the oviduct or after the mammal gastrula has passed into the uterus it is converted into the globular vesicle which is shown externally in Figure 1.106, and in section in Figure 1.107. The thick, outer, structureless envelope that encloses it is the original ovolemma or zona pellucida, modified, and clothed with a layer of albumin that has been deposited on the outside. From this stage the envelope is called the external membrane, the primary chorion or prochorion (a). The real wall of the vesicle enclosed by it consists of a simple layer of ectodermic cells (b), which are flattened by mutual pressure, and generally hexagonal; a light nucleus shines through their fine–grained protoplasm (Figure 1.108). At one part (c) inside this hollow ball we find a circular disc, formed of darker, softer, and rounder cells, the dark–grained entodermic cells (Figure 1.109).

(FIGURE 1.108. Four entodermic cells from the embryonic vesicle of the rabbit.

FIGURE 1.109. Two entodermic cells from the embryonic vesicle of the rabbit.)

The characteristic embryonic form that the developing mammal now exhibits has up to the present usually been called the "blastula" (Bischoff), "sac–shaped embryo" (Baer), "vesicular embryo" (vesicula blastodermica, or, briefly, blastosphaera). The wall of the hollow vesicle, which consists of a single layer of cells, was called the "blastoderm," and was supposed to be equivalent to the cell–layer of the same name that forms the wall of the real blastula of the amphioxus and many of the invertebrates (such as Monoxenia, Figure 1.29 F, G). Formerly this real blastula was generally believed to be equivalent to the embryonic vesicle of the mammal. However, this is by no means the case. What is called the "blastula" of the mammal and the real blastula of the amphioxus and many of the invertebrates are totally different embryonic structures. The latter (blastula) is palingenetic, and precedes the formation of the gastrula. The former (blastodermic vesicle) is cenogenetic, and follows gastrulation. The globular wall of the blastula is a real blastoderm, and consists of homogeneous (blastodermic) cells; it is not yet differentiated into the two primary germinal layers. But the globular wall of the mammal vesicle is the differentiated ectoderm, and at one point in it we find a circular disk of quite different cells—the entoderm. The round cavity, filled with fluid, inside the real blastula is the segmentation–cavity. But the similar cavity within the mammal vesicle is the yelk–sac cavity, which is connected with the incipient gut–cavity. This primitive gut–cavity passes directly into the segmentation–cavity in the mammals, in consequence of the peculiar cenogenetic changes in their gastrulation, which we have considered previously (Chapter 1.9). For these reasons it is very necessary to recognise the secondary embryonic vesicle in the mammal (gastrocystis or blastocystis) as a characteristic structure peculiar to this class, and distinguish it carefully from the primary blastula of the amphioxus and the invertebrates.

(FIGURE 1.110. Ovum of a rabbit from the uterus, one sixth of an inch in diameter. The embryonic vesicle (b) has withdrawn a little from the smooth ovolemma (a). In the middle of the ovolemma we see the round germinal disk (blastodiscus, c), at the edge of which (at d) the inner layer of the embryonic vesicle is already beginning to expand. (Figures 1.110 to 1.114 from Bischoff.)

FIGURE 1.111. The same ovum, seen in profile. Letters as in Figure 1.110.

FIGURE 1.112. Ovum of a rabbit from the uterus, one–fourth of an inch in diameter. The blastoderm is already for the most part two–layered (b). The ovolemma, or outer envelope, is tufted (a).

FIGURE 1.113. The same ovum, seen in profile. Letters as in Figure 1.112.

FIGURE 1.114. Ovum of a rabbit from the uterus, one–third of an inch in diameter. The embryonic vesicle is now nearly everywhere two–layered (k) only remaining one–layered below (at d).

FIGURE 1.115. Round germinative area of the rabbit, divided into the central light area (area pellucida) and the peripheral dark area (area opaca). The light area seems darker on account of the dark ground appearing through it.)

The small, circular, whitish, and opaque spot which the gastric disk (Figure 1.106) forms at a certain part of the surface of the clear and transparent embryonic vesicle has long been known to science, and compared to the germinal disk of the birds and reptiles. Sometimes it has been called the germinal disk, sometimes the germinal spot, and usually the germinative area. From the area the further development of the embryo proceeds. However, the larger part of the embryonic vesicle of the mammal is not directly used for building up the later body, but for the construction of the temporary umbilical vesicle. The embryo separates from this in proportion as it grows at its expense; the two are only connected by the yelk–duct (the stalk of the yelk–sac), and this maintains the direct communication between the cavity of the umbilical vesicle and the forming visceral cavity (Figure 1.105).

The germinative area or gastric disk of the animal consists at first (like the germinal disk of birds and reptiles) merely of the two primary germinal layers, the ectoderm and entoderm. But soon there appears in the middle of the circular disk between the two a third stratum of cells, the rudiment of the middle layer or fibrous layer (mesoderm). This middle germinal layer consists from the first, as we have seen in Chapter 1.10, of two separate epithelial plates, the two layers of the coelom–pouches (parietal and visceral). However, in all the amniotes (on account of the large formation of yelk) these thin middle plates are so firmly pressed together that they seem to represent a single layer. It is thus peculiar to the amniotes that the middle of the germinative area is composed of four germinal layers, the two limiting (or primary) layers and the middle layers between them (Figures 1.96 and 1.97). These four secondary germinal layers can be clearly distinguished as soon as what is called the sickle–groove (or "embryonic sickle") is seen at the hind border of the germinative area. At the borders, however, the germinative area of the mammal only consists of two layers. The rest of the wall of the embryonic vesicle consists at first (but only for a short time in most of the mammals) of a single layer, the outer germinal layer.

(FIGURE 1.116. Oval area, with the opaque whitish border of the dark area without.)

From this stage, however, the whole wall of the embryonic vesicle becomes two–layered. The middle of the germinative area is much thickened by the growth of the cells of the middle layers, and the inner layer expands at the same time, and increases at the border of the disk all round. Lying close on the outer layer throughout, it grows over its inner surface at all points, covers first the upper and then the lower hemisphere, and at last closes in the middle of the inner layer (Figures 1.110 to 1.114). The wall of the embryonic vesicle now consists throughout of two layers of cells, the ectoderm without and the entoderm within. It is only in the centre of the circular area, which becomes thicker and thicker through the growth of the middle layers, that it is made up of all four layers. At the same time, small structureless tufts or warts are deposited on the surface of the outer ovolemma or prochorion, which has been raised above the embryonic vesicle (Figures 1.112 to 1.114 a).

(FIGURE 1.117. Oval germinal disk of the rabbit, magnified about ten times. As the delicate, half–transparent disk lies on a black ground, the pellucid area looks like a dark ring, and the opaque area (lying outside it) like a white ring. The oval shield in the centre also looks whitish, and in its axis we see the dark medullary groove. (From Bischoff.))

We may now disregard both the outer ovolemma and the greater part of the vesicle, and concentrate our attention on the germinative area and the four–layered embryonic disk. It is here alone that we find the important changes which lead to the differentiation of the first organs. It is immaterial whether we examine the germinative area of the mammal (the rabbit, for instance) or the germinal disk of a bird or a reptile (such as a lizard or tortoise). The embryonic processes we are now going to consider are essentially the same in all members of the three higher classes of vertebrates which we call the amniotes. Man is found to agree in this respect with the rabbit, dog, ox, etc.; and in all these animals the germinative area undergoes essentially the same changes as in the birds and reptiles. They are most frequently and accurately studied in the chick, because we can have incubated hens' eggs in any quantity at any stage of development. Moreover, the round germinal disk of the chick passes immediately after the beginning of incubation (within a few hours) from the two–layered to the four–layered stage, the two–layered mesoderm developing from the median primitive groove between the ectoderm and entoderm (Figures 1.82 to 1.95).

The first change in the round germinal disk of the chick is that the cells at its edges multiply more briskly, and form darker nuclei in their protoplasm. This gives rise to a dark ring, more or less sharply set off from the lighter centre of the germinal disk (Figure 1.115). From this point the latter takes the name of the "light area" (area pellucida), and the darker ring is called the "dark area" (area opaca). (In a strong light, as in Figures 1.115 to 1.117, the light area seems dark, because the dark ground is seen through it; and the dark area seems whiter). The circular shape of the area now changes into elliptic, and then immediately into oval (Figures 1.116 and 1.117). One end seems to be broader and blunter, the other narrower and more pointed; the former corresponds to the anterior and the latter to the posterior section of the subsequent body. At the same time, we can already trace the characteristic bilateral form of the body, the antithesis of right and left, before and behind. This will be made clearer by the "primitive streak," which appears at the posterior end.

(FIGURE 1.118. Pear–shaped germinal shield of the rabbit (eight days old), magnified twenty times. rf medullary groove. pr primitive groove (primitive mouth). (From Kolliker.)

FIGURE 1.119. Median longitudinal section of the gastrula of four vertebrates. (From Rabl.) A discogastrula of a shark (Pristiurus). B amphigastrula of a sturgeon (Accipenser). C amphigastrula of an amphibium (Triton). D epigastrula of an amniote (diagram). a ventral, b dorsal lip of the primitive mouth.)

At an early stage an opaque spot is seen in the middle of the clear germinative area, and this also passes from a circular to an oval shape. At first this shield–shaped marking is very delicate and barely perceptible; but it soon becomes clearer, and now stands out as an oval shield, surrounded by two rings or areas (Figure 1.117). The inner and brighter ring is the remainder of the pellucid area, and the dark outer ring the remainder of the opaque area; the opaque shield–like spot itself is the first rudiment of the dorsal part of the embryo. We give it briefly the name of embryonic shield or dorsal shield. In most works this embryonic shield is described as "the first rudiment or trace of the embryo," or "primitive embryo." But this is wrong, though it rests on the authority of Baer and Bischoff. As a matter of fact, we already have the embryo in the stem–cell, the gastrula, and all the subsequent stages. The embryonic shield is simply the first rudiment of the dorsal part, which is the earliest to develop. As the older names of "embryonic rudiment" and "germinative area" are used in many different senses—and this has led to a fatal confusion in embryonic literature—we must explain very clearly the real significance of these important embryonic parts of the amniote. It will be useful to do so in a series of formal principles:—

1. The so–called "first trace of the embryo" in the amniotes, or the embryonic shield, in the centre of the pellucid area, consists merely of an early differentiation and formation of the middle dorsal parts.

2. Hence the best name for it is "the dorsal shield," as I proposed long ago.

3. The germinative area, in which the first embryonic blood–vessels appear at an early stage, is not opposed as an external area to the "embryo proper," but is a part of it.

4. In the same way, the yelk–sac or the umbilical vesicle is not a foreign external appendage of the embryo, but an outlying part of its primitive gut.

5. The dorsal shield gradually separates from the germinative area and the yelk–sac, its edges growing downwards and folding together to form ventral plates.

6. The yelk–sac and vessels of the germinative area, which soon spread over its whole surface, are, therefore, real embryonic organs, or temporary parts of the embryo, and have a transitory importance in connection with the nutrition of the growing later body; the latter may be called the "permanent body" in contrast to them.

The relation of these cenogenetic features of the amniotes to the palingenetic structures of the older non–amniotic vertebrates may be expressed in the following theses: The original gastrula, which completely passes into the embryonic body in the acrania, cyclostoma, and amphibia, is early divided into two parts in the amniotes—the embryonic shield, which represents the dorsal outline of the permanent body; and the temporary embryonic organs of the germinative area and its blood–vessels, which soon grow over the whole of the yelk–sac. The differences which we find in the various classes of the vertebrate stem in these important particulars can only be fully understood when we bear in mind their phylogenetic relations on the one hand, and, on the other, the cenogenetic modifications of structure that have been brought about by changes in the rearing of the young and the variation in the mass of the food–yelk.

We have already described in Chapter 1.9 the changes which this increase and decrease of the nutritive yelk causes in the form of the gastrula, and especially in the situation and shape of the primitive mouth. The primitive mouth or prostoma is originally a simple round aperture at the lower pole of the long axis; its dorsal lip is above and ventral lip below. In the amphioxus this primitive mouth is a little eccentric, or shifted to the dorsal side (Figure 1.39). The aperture increases with the growth of the food–yelk in the cyclostoma and ganoids; in the sturgeon it lies almost on the equator of the round ovum, the ventral lip (a) in front and the dorsal lip (b) behind (Figure 1.119 b). In the wide–mouthed, circular discoid gastrula of the selachii or primitive fishes, which spreads quite flat on the large food–yelk, the anterior semi–circle of the border of the disk is the ventral, and the posterior semicircle the dorsal lip (Figure 1.119 A). The amphiblastic amphibia are directly connected with their earlier fish–ancestors, the dipneusts and ganoids, and further the oldest selachii (Cestracion); they have retained their total unequal segmentation, and their small primitive mouth (Figure 1.119 C, ab), blocked up by the yelk–stopper, lies at the limit of the dorsal and ventral surface of the embryo (at the lower pole of its equatorial axis), and there again has an upper dorsal and a lower ventral lip (a, b). The formation of a large food–yelk followed again in the stem–forms of the amniotes, the protamniotes or proreptilia, descended from the amphibia (Figure 1.119 D). But here the accumulation of the food–yelk took place only in the ventral wall of the primitive–gut, so that the narrow primitive mouth lying behind was forced upwards, and came to lie on the back of the discoid "epigastrula" in the shape of the "primitive groove"; thus (in contrast to the case of the selachii, Figure 1.119 A) the dorsal lip (b) had to be in front, and the ventral lip (a) behind (Figure 1.119 D). This feature was transmitted to all the amniotes, whether they retained the large food–yelk (reptiles, birds, and monotremes), or lost it by atrophy (the viviparous mammals).

This phylogenetic explanation of gastrulation and coelomation, and the comparative study of them in the various vertebrates, throw a clear and full light on many ontogenetic phenomena, as to which the most obscure and confused opinions were prevalent thirty years ago. In this we see especially the high scientific value of the biogenetic law and the careful separation of palingenetic from cenogenetic processes. To the opponents of this law the real explanation of these remarkable phenomena is impossible. Here, and in every other part of embryology, the true key to the solution lies in phylogeny.

CHAPTER 1.13. DORSAL BODY AND VENTRAL BODY.

The earliest stages of the human embryo are, for the reasons already given, either quite unknown or only imperfectly known to us. But as the subsequent embryonic forms in man behave and develop just as they do in all the other mammals, there cannot be the slightest doubt that the preceding stages also are similar. We have been able to see in the coelomula of the human embryo (Figure 1.97), by transverse sections through its primitive mouth, that its two coelom–pouches are developed in just the same way as in the rabbit (Figure 1.96); moreover, the peculiar course of the gastrulation is just the same.

(FIGURE 1.120. Embryonic vesicle of a seven–days–old rabbit with oval embryonic shield (ag). A seen from above, B from the side. (From Kolliker.) ag dorsal shield or embryonic spot. In B the upper half of the vesicle is made up of the two primary germinal layers, the lower (up to ge) only from the outer layer.)

The germinative area forms in the human embryo in the same way as in the other mammals, and in the middle part of this we have the embryonic shield, the purport of which we considered in Chapter 1.12. The next changes in the embryonic disk, or the "embryonic spot," take place in corresponding fashion. These are the changes we are now going to consider more closely.

The chief part of the oval embryonic shield is at first the narrow hinder end; it is in the middle line of this that the primitive streak appears (Figure 1.121 ps). The narrow longitudinal groove in it—the so–called "primitive groove"—is, as we have seen, the primitive mouth of the gastrula. In the gastrula–embryos of the mammals, which are much modified cenogenetically, this cleft–shaped prostoma is lengthened so much that it soon traverses the whole of the hinder half of the dorsal shield; as we find in a rabbit embryo of six to eight days (Figure 1.122 pr). The two swollen parallel borders that limit this median furrow are the side lips of the primitive mouth, right and left. In this way the bilateral–symmetrical type of the vertebrate becomes pronounced. The subsequent head of the amniote is developed from the broader and rounder fore–half of the dorsal shield.

In this fore–half of the dorsal shield a median furrow quickly makes its appearance (Figure 1.123 rf). This is the broader dorsal furrow or medullary groove, the first beginning of the central nervous system. The two parallel dorsal or medullary swellings that enclose it grow together over it afterwards, and form the medullary tube. As is seen in transverse sections, it is formed only of the outer germinal layer (Figures 1.95 and 1.136). The lips of the primitive mouth, however, lie, as we know, at the important point where the outer layer bends over the inner, and from which the two coelom pouches grow between the primary germinal layers.

(FIGURE 1.121. Oval embryonic shield of the rabbit (A of six days eighteen hours, B of eight days). (From Kolliker.) ps primitive streak, pr primitive groove, arg area germinalis, sw sickle–shaped germinal growth.

FIGURE 1.122. Dorsal shield (ag) and germinative area of a rabbit–embryo of eight days. (From Kolliker.) pr primitive groove, rf dorsal furrow.

FIGURE 1.123. Embryonic shield of a rabbit of eight days. (From Van Beneden.) pr primitive groove, cn canalis neurentericus, nk nodus neurentericus (or "Hensen's ganglion"), kf head–process (chorda).

FIGURE 1.124. Longitudinal section of the coelomula of amphioxus (from the left). i entoderm, d primitive gut, cn medullary duct, n nerve tube, m mesoderm, s first primitive segment, c coelom–pouches. (From Hatschek.))

Thus the median primitive furrow (pr) in the hind–half and the median medullary furrow (rf) in the fore–half of the oval shield are totally different structures, although the latter seems to a superficial observer to be merely the forward continuation of the former. Hence they were formerly always confused. This error was the more pardonable as immediately afterwards the two grooves do actually pass into each other in a very remarkable way. The point of transition is the remarkable neurenteric canal (Figure 1.124 cn). But the direct connection which is thus established does not last long; the two are soon definitely separated by a partition.

The enigmatic neurenteric canal is a very old embryonic organ, and of great phylogenetic interest, because it arises in the same way in all the chordonia (both tunicates and vertebrates). In every case it touches or embraces like an arch the posterior end of the chorda, which has been developed here in front out of the middle line of the primitive gut (between the two coelom–folds of the sickle groove) ("head–process," Figure 1.123 kf). These very ancient and strictly hereditary structures, which have no physiological significance to–day, deserve (as "rudimentary organs") our closest attention. The tenacity with which the useless neurenteric canal has been transmitted down to man through the whole series of vertebrates is of equal interest for the theory of descent in general, and the phylogeny of the chordonia in particular.

The connection which the neurenteric canal (Figure 1.123 cn) establishes between the dorsal nerve–tube (n) and the ventral gut–tube (d) is seen very plainly in the amphioxus in a longitudinal section of the coelomula, as soon as the primitive mouth is completely closed at its hinder end. The medullary tube has still at this stage an opening at the forward end, the neuroporus (Figure 1.83 np). This opening also is afterwards closed. There are then two completely closed canals over each other—the medullary tube above and the gastric tube below, the two being separated by the chorda. The same features as in the acrania are exhibited by the related tunicates, the ascidiae.

Again, we find the neurenteric canal in just the same form and situation in the amphibia. A longitudinal section of a young tadpole (Figure 1.125) shows how we may penetrate from the still open primitive mouth (x) either into the wide primitive gut–cavity (al) or the narrow overlying nerve–tube. A little later, when the primitive mouth is closed, the narrow neurenteric canal (Figure 1.126 ne) represents the arched connection between the dorsal medullary canal (mc) and the ventral gastric canal.

(FIGURE 1.125. Longitudinal section of the chordula of a frog. (From Balfour.) nc nerve–tube, x canalis neurentericus, al alimentary canal, yk yelk–cells, m mesoderm.

FIGURE 1.126. Longitudinal section of a frog–embryo. (From Goette.) m mouth, l liver, an anus, ne canalis neurentericus, mc medullary–tube, pn pineal body (epiphysis), ch chorda.

FIGURES 1.127 AND 1.128. Dorsal shield of the chick. (From Balfour.) The medullary furrow (me), which is not yet visible in Figure 1.130, encloses with its hinder end the fore end of the primitive groove (pr) in Figure 1.131.)

In the amniotes this original curved form of the neurenteric canal cannot be found at first, because here the primitive mouth travels completely over to the dorsal surface of the gastrula, and is converted into the longitudinal furrow we call the primitive groove. Hence the primitive groove (Figure 1.128 pr), examined from above, appears to be the straight continuation of the fore–lying and younger medullary furrow (me). The divergent hind legs of the latter embrace the anterior end of the former. Afterwards we have the complete closing of the primitive mouth, the dorsal swellings joining to form the medullary tube and growing over it. The neurenteric canal then leads directly, in the shape of a narrow arch–shaped tube (Figure 1.129 ne), from the medullary tube (sp) to the gastric tube (pag). Directly in front of it is the latter end of the chorda (cli).

While these important processes are taking place in the axial part of the dorsal shield, its external form also is changing. The oval form (Figure 1.117) becomes like the sole of a shoe or sandal, lyre–shaped or finger–biscuit shaped (Figure 1.130). The middle third does not grow in width as quickly as the posterior, and still less than the anterior third; thus the shape of the permanent body becomes somewhat narrow at the waist. At the same time, the oval form of the germinative area returns to a circular shape, and the inner pellucid area separates more clearly from the opaque outer area (Figure 1.131 a). The completion of the circle in the area marks the limit of the formation of blood–vessels in the mesoderm.

(FIGURE 1.129. Longitudinal section of the hinder end of a chick. (From Balfour.) sp medullary tube, connected with the terminal gut (pag) by the neurenteric canal (ne), ch chorda, pr neurenteric (or Hensen's) ganglion, al allantois, ep ectoderm, hy entoderm, so parietal layer, sp visceral layer, an anus–pit, am amnion.)

The characteristic sandal–shape of the dorsal shield, which is determined by the narrowness of the middle part, and which is compared to a violin, lyre, or shoe–sole, persists for a long time in all the amniotes. All mammals, birds, and reptiles have substantially the same construction at this stage, and even for a longer or shorter period after the division of the primitive segments into the coelom–folds has begun (Figure 1.132). The human embryonic shield assumes the sandal–form in the second week of development; towards the end of the week our sole–shaped embryo has a length of about one–twelfth of an inch (Figure 1.133).

The complete bilateral symmetry of the vertebrate body is very early indicated in the oval form of the embryonic shield (Figure 1.117) by the median primitive streak; in the sandal–form it is even more pronounced (Figures 1.131 to 1.135). In the lateral parts of the embryonic shield a darker central and a lighter peripheral zone become more obvious; the former is called the stem–zone (Figure 1.134 stz), and the latter the parietal zone (pz); from the first we get the dorsal and from the second the ventral half of the body–wall. The stem–zone of the amniote embryo would be called more appropriately the dorsal zone or dorsal shield; from it develops the whole of the dorsal half of the later body (or permanent body)—that is to say, the dorsal body (episoma). Again, it would be better to call the "parietal zone" the ventral zone or ventral shield; from it develop the ventral "lateral plates," which afterwards separate from the embryonic vesicle and form the ventral body (hyposoma)—that is to say, the ventral half of the permanent body, together with the body–cavity and the gastric canal that it encloses.

(FIGURE 1.130. Germinal area or germinal disk of the rabbit, with sole–shaped embryonic shield, magnified about ten times. The clear circular field (d) is the opaque area. The pellucid area (c) is lyre–shaped, like the embryonic shield itself (b). In its axis is seen the dorsal furrow or medullary furrow (a). (From Bischoff.))

The sole–shaped germinal shields of all the amniotes are still, at the stage of construction which Figure 1.134 illustrates in the rabbit and Figure 1.135 in the opossum, so like each other that we can either not distinguish them at all or only by means of quite subordinate peculiarities in the size of the various parts. Moreover, the human sandal–shaped embryo cannot at this stage be distinguished from those of other mammals, and it particularly resembles that of the rabbit. On the other hand, the outer form of these flat sandal–shaped embryos is very different from the corresponding form of the lower animals, especially the acrania (amphioxus). Nevertheless, the body is just the same in the essential features of its structure as that we find in the chordula of the latter (Figures 1.83 to 1.86), and in the embryonic forms which immediately develop from it. The striking external difference is here again due to the fact that in the palingenetic embryos of the amphioxus (Figures 1.83 and 1.84) and the amphibia (Figures 1.85 and 1.86) the gut–wall and body–wall form closed tubes from the first, whereas in the cenogenetic embryos of the amniotes they are forced to expand leaf–wise on the surface owing to the great extension of the food–yelk.

(FIGURE 1.131. Embryo of the opossum, sixty hours old, one–sixth of an inch in diameter. (From Selenka) b the globular embryonic vesicle, a the round germinative area, b limit of the ventral plates, r dorsal shield, v its fore part, u the first primitive segment, ch chorda, chr its fore–end, pr primitive groove (or mouth).

FIGURE 1.132. Sandal–shaped embryonic shield of a rabbit of eight days, with the fore part of the germinative area (ao opaque, ap pellucid area). (From Kolliker.) rf dorsal furrow, in the middle of the medullary plate, h, pr primitive groove (mouth), stz dorsal (stem) zone, pz ventral (parietal) zone. In the narrow middle part the first three primitive segments may be seen.)

It is all the more notable that the early separation of dorsal and ventral halves takes place in the same rigidly hereditary fashion in all the vertebrates. In both the acrania and the craniota the dorsal body is about this period separated from the ventral body. In the middle part of the body this division has already taken place by the construction of the chorda between the dorsal nerve–tube and the ventral canal. But in the outer or lateral part of the body it is only brought about by the division of the coelom–pouches into two sections—a dorsal episomite (dorsal segment or provertebra) and a ventral hyposomite (or ventral segment) by a frontal constriction. In the amphioxus each of the former makes a muscular pouch, and each of the latter a sex–pouch or gonad.

These important processes of differentiation in the mesoderm, which we will consider more closely in the next chapter, proceed step by step with interesting changes in the ectoderm, while the entoderm changes little at first. We can study these processes best in transverse sections, made vertically to the surface through the sole–shaped embryonic shield. Such a transverse section of a chick embryo, at the end of the first day of incubation, shows the gut–gland layer as a very simple epithelium, which is spread like a leaf over the outer surface of the food–yelk (Figure 1.92). The chorda (ch) has separated from the dorsal middle line of the entoderm; to the right and left of it are the two halves of the mesoderm, or the two coelom–folds. A narrow cleft in the latter indicates the body–cavity (uwh); this separates the two plates of the coelom–pouches, the lower (visceral) and upper (parietal). The broad dorsal furrow (Rf) formed by the medullary plate (m) is still wide open, but is divided from the lateral horn–plate (h) by the parallel medullary swellings, which eventually close.

(FIGURE 1.133. Human embryo at the sandal–stage, one–twelfth of an inch long, from the end of the second week, magnified twenty–five times. (From Count Spee.)

FIGURE 1.134. Sandal–shaped embryonic shield of a rabbit of nine days. (From Kolliker.) (Back view from above.) stz stem–zone or dorsal shield (with eight pairs of primitive segments), pz parietal or ventral zone, ap pellucid area, af amnion–fold, h heart, ph pericardial cavity, vo omphalo–mesenteric vein, ab eye–vesicles, vh fore brain, mh middle brain, hh hind brain, uw primitive segments (or vertebrae).)

During these processes important changes are taking place in the outer germinal layer (the "skin–sense layer"). The continued rise and growth of the dorsal swellings causes their higher parts to bend together at their free borders, approach nearer and nearer (Figure 1.136 w), and finally unite. Thus in the end we get from the open dorsal furrow, the upper cleft of which becomes narrower and narrower, a closed cylindrical tube (Figure 1.137 mr). This tube is of the utmost importance; it is the beginning of the central nervous system, the brain and spinal marrow, the medullary tube. This embryonic fact was formerly looked upon as very mysterious. We shall see presently that in the light of the theory of descent it is a thoroughly natural process. The phylogenetic explanation of it is that the central nervous system is the organ by means of which all intercourse with the outer world, all psychic action and sense–perception, are accomplished; hence it was bound to develop originally from the outer and upper surface of the body, or from the outer skin. The medullary tube afterwards separates completely from the outer germinal layer, and is surrounded by the middle parts of the provertebrae and forced inwards (Figure 1.146). The remaining portion of the skin–sense layer (Figure 1.93 h) is now called the horn–plate or horn–layer, because from it is developed the whole of the outer skin or epidermis, with all its horny appendages (nails, hair, etc.).

(FIGURE 1.135. Sandal–shaped embryonic shield of an opossum (Didelphys), three days old. (From Selenka.) (Back view from above.) stz stem–zone or dorsal shield (with eight pairs of primitive segments), pz parietal or ventral zone, ap pellucid area, ao opaque area, hh halves of the heart, v fore–end, h hind–end. In the median line we see the chorda (ch) through the transparent medullary tube (m). u primitive segment, pr primitive streak (or primitive mouth).)

A totally different organ, the prorenal (primitive kidney) duct (ung), is found to be developed at an early stage from the ectoderm. This is originally a quite simple, tube–shaped, lengthy duct, or straight canal, which runs from front to rear at each side of the provertebrae (on the outer side, Figure 1.93 ung). It originates, it seems, out of the horn–plate at the side of the medullary tube, in the gap that we find between the provertebral and the lateral plates. The prorenal duct is visible in this gap even at the time of the severance of the medullary tube from the horn–plate. Other observers think that the first trace of it does not come from the skin–sense layer, but the skin–fibre layer.

The inner germinal layer, or the gut–fibre layer (Figure 1.93 dd), remains unchanged during these processes. A little later, however, it shows a quite flat, groove–like depression in the middle line of the embryonic shield, directly under the chorda. This depression is called the gastric groove or furrow. This at once indicates the future lot of this germinal layer. As this ventral groove gradually deepens, and its lower edges bend towards each other, it is formed into a closed tube, the alimentary canal, in the same way as the medullary groove grows into the medullary tube. The gut–fibre layer (Figure 1.137 f), which lies on the gut–gland layer (d), naturally follows it in its folding. Moreover, the incipient gut–wall consists from the first of two layers, internally the gut–gland layer and externally the gut–fibre layer.

The formation of the alimentary canal resembles that of the medullary tube to this extent—in both cases a straight groove or furrow arises first of all in the middle line of a flat layer. The edges of this furrow then bend towards each other, and join to form a tube (Figure 1.137). But the two processes are really very different. The medullary tube closes in its whole length, and forms a cylindrical tube, whereas the alimentary canal remains open in the middle, and its cavity continues for a long time in connection with the cavity of the embryonic vesicle. The open connection between the two cavities is only closed at a very late stage, by the construction of the navel. The closing of the medullary tube is effected from both sides, the edges of the groove joining together from right and left. But the closing of the alimentary canal is not only effected from right and left, but also from front and rear, the edges of the ventral groove growing together from every side towards the navel. Throughout the three higher classes of vertebrates the whole of this process of the construction of the gut is closely connected with the formation of the navel, or with the separation of the embryo from the yelk–sac or umbilical vesicle.

In order to get a clear idea of this, we must understand carefully the relation of the embryonic shield to the germinative area and the embryonic vesicle. This is done best by a comparison of the five stages which are shown in longitudinal section in Figures 1.138

to 1.142. The embryonic shield (c), which at first projects very slightly over the surface of the germinative area, soon begins to rise higher above it, and to separate from the embryonic vesicle. At this point the embryonic shield, looked at from the dorsal surface, shows still the original simple sandal–shape (Figures 1.133 to 1.135). We do not yet see any trace of articulation into head, neck, trunk, etc., or limbs. But the embryonic shield has increased greatly in thickness, especially in the anterior part. It now has the appearance of a thick, oval swelling, strongly curved over the surface of the germinative area. It begins to sever completely from the embryonic vesicle, with which it is connected at the ventral surface. As this severance proceeds, the back bends more and more; in proportion as the embryo grows the embryonic vesicle decreases, and at last it merely hangs as a small vesicle from the belly of the embryo (Figure 1.142 ds). In consequence of the growth–movements which cause this severance, a groove–shaped depression is formed at the surface of the vesicle, the limiting furrow, which surrounds the vesicle in the shape of a pit, and a circular mound or dam (Figure 1.139 ks) is formed at the outside of this pit by the elevation of the contiguous parts of the germinal vesicle.

(FIGURE 1.136. Transverse section of the embryonic disk of a chick at the end of the first day of incubation, magnified about twenty times. The edges of the medullary plate (m), the medullary swellings (w), which separate the medullary from the horn–plate (h), are bending towards each other. At each side of the chorda (ch) the primitive segment plates (u) have separated from the lateral plates (sp). A gut–gland layer. (From Remak.))

In order to understand clearly this important process, we may compare the embryo to a fortress with its surrounding rampart and trench. The ditch consists of the outer part of the germinative area, and comes to an end at the point where the area passes into the vesicle. The important fold of the middle germinal layer that brings about the formation of the body–cavity spreads beyond the borders of the embryo over the whole germinative area. At first this middle layer reaches as far as the germinative area; the whole of the rest of the embryonic vesicle consists in the beginning only of the two original limiting layers, the outer and inner germinal layers. Hence, as far as the germinative area extends the germinal layer splits into the two plates we have already recognised in it, the outer skin–fibre layer and the inner gut–fibre layer. These two plates diverge considerably, a clear fluid gathering between them (Figure 1.140 am). The inner plate, the gut–fibre layer, remains on the inner layer of the embryonic vesicle (on the gut–gland layer). The outer plate, the skin–fibre layer, lies close on the outer layer of the germinative area, or the skin–sense layer, and separates together with this from the embryonic vesicle. From these two united outer plates is formed a continuous membrane. This is the circular mound that rises higher and higher round the whole embryo, and at last joins above it (Figures 1.139 to 1.142 am). To return to our illustration of the fortress, we must imagine the circular rampart to be extraordinarily high and towering far above the fortress. Its edges bend over like the combs of an overhanging wall of rock that would enclose the fortress; they form a deep hollow, and at last join together above. In the end the fortress lies entirely within the hollow that has been formed by the growth of the edges of this large rampart.

(FIGURE 1.137. Three diagrammatic transverse sections of the embryonic disk of the higher vertebrate, to show the origin of the tubular organs from the bending germinal layers. In Figure A the medullary tube (n) and the alimentary canal (a) are still open grooves. In Figure B the medullary tube (n) and the dorsal wall are closed, but the alimentary canal (a) and the ventral wall are open; the prorenal ducts (u) are cut off from the horn−plate (h) and internally connected with segmental prorenal canals. In Figure C both the medullary tube and the dorsal wall above and the alimentary canal and ventral wall below are closed. All the open grooves have become closed tubes; the primitive kidneys are directed inwards. The letters have the same meaning in all three figures: h skin−sense layer, n medullary tube, u prorenal ducts, x axial rod, s primitive−vertebra, r dorsal wall, b ventral wall, c body−cavity or coeloma, f gut−fibre layer, t primitive artery (aorta), v primitive vein (subintestinal vein), d gut−fibre layer, a alimentary canal.)

As the two outer layers of the germinative area thus rise in a fold about the embryo, and join above it, they come at last to form a spacious sac−like membrane about it. This envelope takes the name of the germinative membrane, or water−membrane, or amnion (Figure 1.142 am). The embryo floats in a watery fluid, which fills the space between the embryo and the amnion, and is called the amniotic fluid (Figures 1.141 and 1.142 ah). We will deal with this remarkable formation and with the allantois later on (Chapter 1.15). In front of the allantois the yelk−sac or umbilical vesicle (ds), the remainder of the original embryonic vesicle, starts from the open belly of the embryo (Figure 1.138 kh). In more advanced embryos, in which the gastric wall and the ventral wall are nearly closed, it hangs out of the navel−opening in the shape of a small vesicle with a stalk (Figures 1.141 and 1.142 ds). The more the embryo grows, the smaller becomes the vitelline (yelk) sac. At first the embryo looks like a small appendage of the large embryonic vesicle. Afterwards it is the yelk−sac, or the remainder of the embryonic vesicle, that seems a small pouch−like appendage of the embryo (Figure 1.142 ds). It ceases to have

195

any significance in the end. The very wide opening, through which the gastric cavity at first communicates with the umbilical vesicle, becomes narrower and narrower, and at last disappears altogether. The navel, the small pit–like depression that we find in the developed man in the middle of the abdominal wall, is the spot at which the remainder of the embryonic vesicle (the umbilical vesicle) originally entered into the ventral cavity, and joined on to the growing gut.

(FIGURES 1.138 TO 1.142. Five diagrammatic longitudinal sections of the maturing mammal embryo and its envelopes. In Figures 1.138 to 1.141 the longitudinal section passes through the sagittal or middle plane of the body, dividing the right and left halves; in Figure 1.142 the embryo is seen from the left side. In Figure 1.138 the tufted it prochorion (dd apostrophe) encloses the germinal vesicle, the wall of which consists of the two primary layers. Between the outer (a) and inner (i) layer the middle layer (m) has been developed in the region of the germinative area. In Figure 1.139 the embryo (e) begins to separate from the embryonic vesicle (ds), while the wall of the amnion–fold rises about it (in front as head–sheath, ks, behind as tail–sheath, ss). In Figure 1.140 the edges of the amniotic fold (am) rise together over the back of the embryo, and form the amniotic cavity (ah); as the embryo separates more completely from the embryonic vesicle (ds) the alimentary canal (dd) is formed, from the hinder end of which the allantois grows (al). In Figure 1.141 the allantois is larger; the yelk–sac (ds) smaller. In Figure 1.142 the embryo shows the gill–clefts and the outline of the two legs; the chorion has formed branching villi (tufts.) In all four figures e = embryo, a outer germinal layer, m middle germinal layer, i inner germinal layer, am amnion (ks head–sheath, ss tail–sheath), ah amniotic cavity, as amniotic sheath of the umbilical cord, kh embryonic vesicle, ds yelk–sac (umbilical vesicle), dg vitelline duct, df gut–fibre layer, dd gut–gland layer, al allantois, vl = hh place of heart, d vitelline membrane (ovolemma or prochorion), d apostrophe tufts or villi of same, sh serous membrane (serolemma), sz tufts of same, ch chorion, chz tufts or villi, st terminal vein, r pericoelom or serocoelom (the space, filled with fluid, between the amnion and chorion). (From Kolliker.))

The origin of the navel coincides with the complete closing of the external ventral wall. In the amniotes the ventral wall originates in the same way as the dorsal wall. Both are formed substantially from the skin–fibre layer, and externally covered with the horn–plate, the border section of the skin–sense layer. Both come into existence by the conversion of the four flat germinal layers of the embryonic shield into a double tube by folding from opposite directions; above, at the back, we have the vertebral canal which encloses the medullary tube, and below, at the belly, the wall of the body−cavity which contains the alimentary canal (Figure 1.137).

We will consider the formation of the dorsal wall first, and that of the ventral wall afterwards (Figures 1.143 to 1.147). In the middle of the dorsal surface of the embryo there is originally, as we already know, the medullary (mr) tube directly underneath the horn−plate (h), from the middle part of which it has been developed. Later, however, the provertebral plates (uw) grow over from the right and left between these originally connected parts (Figures 1.145 and 1.146). The upper and inner edges of the two provertebral plates push between the horn−plate and medullary tube, force them away from each other, and finally join between them in a seam that corresponds to the middle line of the back. The coalescence of these two dorsal plates and the closing in the middle of the dorsal wall take place in the same way as the medullary tube, which is henceforth enclosed by the vertebral tube. Thus is formed the dorsal wall, and the medullary tube takes up a position inside the body. In the same way the provertebral mass grows afterwards round the chorda, and forms the vertebral column. Below this the inner and outer edge of the provertebral plate splits on each side into two horizontal plates, of which the upper pushes between the chorda and medullary tube, and the lower between the chorda and gastric tube. As the plates meet from both sides above and below the chorda, they completely enclose it, and so form the tubular, outer chord−sheath, the sheath from which the vertebral column is formed (perichorda, Figure 1.137 C, s; Figures 1.145 uwh, 1.146).

(FIGURES 1.143 TO 1.146. Transverse sections of embryos (of chicks). Figure 1.143 of the second, Figure 1.144 of the third, Figure 1.145 of the fourth, and Figure 1.146 of the fifth day of incubation. Figures 1.143 to 1.145 from Kolliker, magnified about 100 times; Figure 1.146 from Remak, magnified about twenty times. h horn−plate, mr medullary tube, ung prorenal duct, un prorenal vesicles, hp skin−fibre layer, m = mu = mp muscle−plate, uw provertebral plate (wh cutaneous rudiment of the body of the vertebra, wb of the arch of the vertebra, wq the rib or transverse continuation), uwh provertebral cavity, ch axial rod or chorda, sh chorda−sheath, bh ventral wall, g hind and v fore root of the spinal nerves, a = af = am amniotic fold, p body−cavity or coeloma, df gut−fibre layer, ao primitive aortas, sa secondary aorta, vc cardinal veins, d = dd gut−gland layer, dr gastric groove. In Figure 1.143 the larger part of the right half, in Figure 1.144 the larger part of the left half, of the section is omitted. Of the yelk−sac or remainder of the embryonic vesicle only a small piece of the wall is indicated below.)

We find in the construction of the ventral wall precisely the same processes as in the formation of the dorsal wall (Figure 1.137 B, Figure 1.144 hp, Figure 1.146 bh). It is formed on the flat embryonic shield of the amniotes from the upper plates of the parietal zone. The right and left parietal plates bend downwards towards each other, and grow round the gut in the same way as the gut itself closes. The outer part of the lateral plates forms the ventral wall or the lower wall of the body, the two lateral plates bending considerably on the inner side of the amniotic fold, and growing towards each other from right and left. While the alimentary canal is closing, the body–wall also closes on all sides. Hence the ventral wall, which encloses the whole ventral cavity below, consists of two parts, two lateral plates that bend towards each other. These approach each other all along, and at last meet at the navel. We ought, therefore, really to distinguish two navels, an inner and an outer one. The internal or intestinal navel is the definitive point of the closing of the gut wall, which puts an end to the open communication between the ventral cavity and the cavity of the yelk–sac (Figure 1.105). The external navel in the skin is the definitive point of the closing of the ventral wall; this is visible in the developed body as a small depression.

(FIGURE 1.147. Median longitudinal section of the embryo of a chick (fifth day of incubation), seen from the right side (head to the right, tail to the left). Dorsal body dark, with convex outline. d gut, o mouth, a anus, l lungs, h liver, g mesentery, v auricle of the heart, k ventricle of the heart, b arch of the arteries, t aorta, c yelk–sac, m vitelline (yelk) duct, u allantois, r pedicle (stalk) of the allantois, n amnion, w amniotic cavity, s serous membrane. (From Baer.))

With the formation of the internal navel and the closing of the alimentary canal is connected the formation of two cavities, which we call the capital and the pelvic sections of the visceral cavity. As the embryonic shield lies flat on the wall of the embryonic vesicle at first, and only gradually separates from it, its fore and hind ends are independent in the beginning; on the other hand, the middle part of the ventral surface is connected with the yelk–sac by means of the vitelline or umbilical duct (Figure 1.147 m). This leads to a notable curving of the dorsal surface; the head–end bends downwards towards the breast and the tail–end towards the belly. We see this very clearly in the excellent old diagrammatic illustration given by Baer (Figure 1.147), a median longitudinal section of the embryo of the chick, in which the dorsal body or episoma is deeply shaded. The embryo seems to be trying to roll up, like a hedgehog protecting itself from its pursuers. This pronounced curve of the back is due to the more rapid growth of the convex dorsal surface, and is directly connected with the severance of the embryo from the yelk–sac. The further bending of the embryo leads to the formation of the "head–cavity" of the gut (Figure 1.148 above D) and a similar one at the tail, known as its "pelvic cavity."

As a result of these processes the embryo attains a shape that may be compared to a wooden shoe, or, better still, to an overturned canoe. Imagine a canoe or boat with both ends rounded and a small covering before and behind; if this canoe is turned upside down, so that the curved keel is uppermost, we have a fair picture of the canoe–shaped embryo (Figure 1.147). The upturned convex keel corresponds to the middle line of the back; the small chamber underneath the fore–deck represents the capital cavity, and the small chamber under the rear–deck the pelvic chamber of the gut (cf. Figure 1.140).

The embryo now, as it were, presses into the outer surface of the embryonic vesicle with its free ends, while it moves away from it with its middle part. As a result of this change the yelk–sac becomes henceforth only a pouch–like outer appendage at the middle of the ventral wall. The ventral appendage, growing smaller and smaller, is afterwards called the umbilical (navel) vesicle. The cavity of the yelk–sac or umbilical vesicle communicates with the corresponding visceral cavity by a wide opening, which gradually contracts into a narrow and long canal, the vitelline (yelk) duct (ductus vitellinus, Figure 1.147 m). Hence, if we were to imagine ourselves in the cavity of the yelk–sac, we could get from it through the yelk–duct into the middle and still wide open part of the alimentary canal. If we were to go forward from there into the head–part of the embryo, we should reach the capital cavity of the gut, the fore–end of which is closed up.

The reader will ask: "Where are the mouth and the anus?" These are not at first present in the embryo. The whole of the primitive gut–cavity is completely closed, and is merely connected in the middle by the vitelline duct with the equally closed cavity of the embryonic vesicle (Figure 1.140). The two later apertures of the alimentary canal—the anus and the mouth—are secondary constructions, formed from the outer skin. In the horn–plate, at the spot where the mouth is found subsequently, a pit–like depression is formed, and this grows deeper and deeper, pushing towards the blind fore–end of the capital cavity; this is the mouth–pit. In the same way, at the spot in the outer skin where the anus is afterwards situated a pit–shaped depression appears, grows deeper and deeper, and approaches the blind hind–end of the pelvic cavity; this is the anus–pit. In the end these pits touch with their deepest and innermost points the two blind ends of the primitive alimentary canal, so that they are now only separated from them by thin membranous partitions. This membrane finally disappears, and henceforth the alimentary canal opens in front at the mouth and in the rear by the anus (Figures 1.141 and 1.147). Hence at first, if we penetrate into these pits from without, we find a partition cutting them off from the cavity of the alimentary canal, which gradually disappears. The formation of mouth and anus is secondary in all the vertebrates.

(FIGURE 1.148. Longitudinal section of the fore half of a chick–embryo at the end of the first day of incubation (seen from the left side). k head–plates, ch chorda. Above it is the blind fore–end of the ventral tube (m); below it the capital cavity of the gut. d gut–gland layer, df gut–fibre layer, h horn plate, hh cavity of the heart, hk heart–capsule, ks head–sheath, kk head–capsule. (From Remak.))

During the important processes which lead to the formation of the navel, and of the intestinal wall and ventral wall, we find a number of other interesting changes taking place in the embryonic shield of the amniotes. These relate chiefly to the prorenal ducts and the first blood–vessels. The prorenal (primitive kidney) ducts, which at first lie quite flat under the horn–plate or epiderm (Figure 1.93 ung), soon back towards each other in consequence of special growth movements (Figures 1.143 to 1.145 ung). They depart more and more from their point of origin, and approach the gut–gland layer. In the end they lie deep in the interior, on either side of the mesentery, underneath the chorda, (Figure 1.145 ung). At the same time, the two primitive aortas change their position (cf. Figures 1.138 to 1.145 ao); they travel inwards underneath the chorda, and there coalesce at last to form a single secondary aorta, which is found under the rudimentary vertebral column (Figure 1.145 ao). The cardinal veins, the first venous blood–vessels, also back towards each other, and eventually unite immediately above the rudimentary kidneys (Figures 1.145 vc, 152 cav). In the same spot, at the inner side of the fore–kidneys, we soon see the first trace of the sexual organs. The most important part of this apparatus (apart from all its appendages) is the ovary in the female and the testicle in the male. Both develop from a small part of the cell–lining of the body–cavity, at the spot where the skin–fibre layer and gut–fibre layer touch. The connection of this embryonic gland with the prorenal ducts, which lie close to it and assume most important relations to it, is only secondary.

(FIGURE 1.149. Longitudinal section of a human embryo of the fourth week, one–fifth of an inch long, magnified fifteen times. Showing: bend of skull, yelk–sac, umbilical cord, terminal gut, rudimentary kidneys, mesoderm, head–gut (with gill–clefts), primitive lungs, liver, stomach, pancreas, mesentery, primitive kidneys, allantoic duct, rectum. (From Kollmann.)

FIGURE 1.150. Transverse section of a human embryo of fourteen days. mr medullary tube, ch chorda. vu umbilical vein, mt myotome, mp middle plate, ug prorenal duct, lh body–cavity, e ectoderm, bh ventral skin, hf skin–fibre layer, df gut–fibre layer. (From Kollmann.)

FIGURE 1.151. Transverse section of a shark–embryo (or young selachius). mr medullary tube, ch chorda, a aorta, d gut, vp principal (or subintestinal) vein, mt myotome, mm muscular mass of the provertebra, mp middle plate, ug prorenal duct, lh body–cavity, e ectoderm of the rudimentary extremities, mz mesenchymic cells, z point where the myotome and nephrotome separate. (From H.E. Ziegler.)

FIGURE 1.152. Transverse section of a duck–embryo with twenty–four primitive segments. (From Balfour.) From a dorsal lateral joint of the medullary tube (spc) the spinal ganglia (spg) grow out between it and the horn–plate. ch chorda, ao double aorta, hy gut–gland layer, sp gut–fibre layer, with blood–vessels in section, ms muscle plate, in the dorsal wall of the myocoel (episomite). Below the cardinal vein (cav) is the prorenal duct (wd) and a segmental prorenal canal (st). The skin–fibre layer of the body–wall (so) is continued in the amniotic fold (am). Between the four secondary germinal layers and the structures formed from them there is formed embryonic connective matter with stellate cells and vascular structures (Hertwig's "mesenchym").)

CHAPTER 1.14. THE ARTICULATION OF THE BODY.*

(* The term articulation is used in this chapter to denote both "segmentation" and "articulation" in the ordinary sense.—Translator.)

The vertebrate stem, to which our race belongs as one of the latest and most advanced outcomes of the natural development of life, is rightly placed at the head of the animal kingdom. This privilege must be accorded to it, not only because man does in point of fact soar far above all other animals, and has been lifted to the position of "lord of creation"; but also because the vertebrate organism far surpasses all the other animal–stems in size, in complexity of structure, and in the advanced character of its functions. From the point of view of both anatomy and physiology, the vertebrate stem outstrips all the other, or invertebrate, animals.

There is only one among the twelve stems of the animal kingdom that can in many respects be compared with the vertebrates, and reaches an equal, if not a greater, importance in many points. This is the stem of the articulates, composed of three classes: 1, the annelids (earth–worms, leeches, and cognate forms); 2, the crustacea (crabs, etc.); 3, the tracheata (spiders, insects, etc.). The stem of the articulates is superior not only to the vertebrates, but to all other animal–stems, in variety of forms, number of species, elaborateness of individuals, and general importance in the economy of nature.

When we have thus declared the vertebrates and the articulates to be the most important and most advanced of the twelve stems of the animal kingdom, the question arises whether this special position is accorded to them on the ground of a peculiarity of organisation that is common to the two. The answer is that this is really the case; it is their segmental or transverse articulation, which we may briefly call metamerism. In all the vertebrates and articulates the developed individual consists of a series of successive members (segments or metamera = "parts"); in the embryo these are called primitive segments or somites. In each of these segments we have a certain group of organs reproduced in the same arrangement, so that we may regard each segment as an individual unity, or a special "individual" subordinated to the entire personality.

The similarity of their segmentation, and the consequent physiological advance in the two stems of the vertebrates and articulates, has led to the assumption of a direct affinity between them, and an attempt to derive the former directly from the latter. The annelids were supposed to be the direct ancestors, not only of the crustacea and tracheata, but also of the vertebrates. We shall see later (Chapter 2.20) that this annelid theory of the vertebrates is entirely wrong, and ignores the most important differences in the organisation of the two stems. The internal articulation of the vertebrates is just as profoundly different from the external metamerism of the articulates as are their skeletal structure, nervous system, vascular system, and so on. The articulation has been developed in a totally different way in the two stems. The unarticulated chordula (Figures 1.83 to 1.86), which we have recognised as one of the chief palingenetic embryonic forms of the vertebrate group, and from which we have inferred the existence of a corresponding ancestral form for all the vertebrates and tunicates, is quite unthinkable as the stem—form of the articulates.

All articulated animals came originally from unarticulated ones. This phylogenetic principle is as firmly established as the ontogenetic fact that every articulated animal—form develops from an unarticulated embryo. But the organisation of the embryo is totally different in the two stems. The chordula—embryo of all the vertebrates is characterised by the dorsal medullary tube, the neurenteric canal, which passes at the primitive mouth into the alimentary canal, and the axial chorda between the two. None of the articulates, either annelids or arthropods (crustacea and tracheata), show any trace of this type of organisation. Moreover, the development of the chief systems of organs proceeds in the opposite way in the two stems. Hence the segmentation must have arisen independently in each. This is not at all surprising; we find analogous cases in the stalk—articulation of the higher plants and in several groups of other animal stems.

The characteristic internal articulation of the vertebrates and its importance in the organisation of the stem are best seen in the study of the skeleton. Its chief and central part, the cartilaginous or bony vertebral column, affords an obvious instance of vertebrate metamerism; it consists of a series of cartilaginous or bony pieces, which have long been known as vertebrae (or spondyli). Each vertebra is directly connected with a special section of the muscular system, the nervous system, the vascular system, etc. Thus most of the "animal organs" take part in this vertebration. But we saw, when we were considering our own vertebrate character (in Chapter 1.11), that the same internal articulation is also found in the lowest primitive vertebrates, the acrania, although here the whole skeleton consists merely of the simple chorda, and is not at all articulated. Hence the articulation does not proceed primarily from the skeleton, but from the muscular system, and is clearly determined by the more advanced swimming—movements of the primitive chordonia—ancestors.

(FIGURES 1.153 TO 1.155. Sole—shaped embryonic disk of the chick, in three successive stages of development, looked at from the dorsal surface, magnified about twenty times, somewhat diagrammatic. Figure 1.153 with six pairs of somites. Brain a simple vesicle (hb). Medullary furrow still wide open from x; greatly widened at z. mp medullary plates, sp lateral plates, y limit of gullet—cavity (sh) and fore—gut (vd). Figure 1.154 with ten pairs of somites. Brain divided into three vesicles: v fore—brain, m middle—brain, h hind—brain, c heart, dv vitelline—veins. Medullary furrow still wide open behind (z). mp medullary plates. Figure 1.155 with sixteen pairs of somites. Brain divided into five vesicles: v fore–brain, z intermediate–brain, m middle–brain, h hind–brain, n after–brain, a optic vesicles, g auditory vesicles, c heart, dv vitelline veins, mp medullary plate, uw primitive vertebra.)

It is, therefore, wrong to describe the first rudimentary segments in the vertebrate embryo as primitive vertebrae or provertebrae; the fact that they have been so called for some time has led to much error and misunderstanding. Hence we shall give the name of "somites" or primitive segments to these so–called "primitive vertebrae." If the latter name is retained at all, it should only be used of the sclerotom—i.e., the small part of the somites from which the later vertebra does actually develop.

Articulation begins in all vertebrates at a very early embryonic stage, and this indicates the considerable phylogenetic age of the process. When the chordula (Figures 1.83 to 1.86) has completed its characteristic composition, often even a little earlier, we find in the amniotes, in the middle of the sole–shaped embryonic shield, several pairs of dark square spots, symmetrically distributed on both sides of the chorda (Figures 1.131 to 1.135). Transverse sections (Figure 1.93 uw) show that they belong to the stem–zone (episoma) of the mesoderm, and are separated from the parietal zone (hyposoma) by the lateral folds; in section they are still quadrangular, almost square, so that they look something like dice. These pairs of "cubes" of the mesoderm are the first traces of the primitive segments or somites, the so–called "protovertebrae." (Figures 1.153 to 1.155 uw).

(FIGURE 1.156. Embryo of the amphioxus, sixteen hours old, seen from the back. (From Hatschek.) d primitive gut, u primitive mouth, p polar cells of the mesoderm, c coelom–pouches, m their first segment, n medullary tube, i entoderm, e ectoderm, s first segment–fold.

FIGURE 1.157. Embryo of the amphioxus, twenty hours old, with five somites. (Right view; for left view see Figure 1.124.) (From Hatschek.) V fore end, H hind end. ak, mk, ik outer, middle, and inner germinal layers; dh alimentary canal, n neural tube, cn canalis neurentericus, ush coelom–pouches (or primitive–segment cavities), us1 first (and foremost) primitive segment.)

Among the mammals the embryos of the marsupials have three pairs of somites (Figure 1.131) after sixty hours, and eight pairs after seventy–two hours (Figure 1.135). They develop more slowly in the embryo of the rabbit; this has three somites on the eighth day (Figure 1.132), and eight somites a day later (Figure 1.134). In the incubated hen's egg the first somites make their appearance thirty hours after incubation begins (Figure 1.153). At the end of the second day the number has risen to sixteen or eighteen (Figure 1.155). The articulation of the stem–zone, to which the somites owe their origin, thus proceeds briskly from front to rear, new transverse constrictions of the "protovertebral plates" forming continuously and successively. The first segment, which is almost half–way down in the embryonic shield of the amniote, is the foremost of all; from this first somite is formed the first cervical vertebra with its muscles and skeletal parts. It follows from this, firstly, that the multiplication of the primitive segments proceeds backwards from the front, with a constant lengthening of the hinder end of the body; and, secondly, that at the beginning of segmentation nearly the whole of the anterior half of the sole–shaped embryonic shield of the amniote belongs to the later head, while the whole of the rest of the body is formed from its hinder half. We are reminded that in the amphioxus (and in our hypothetic primitive vertebrate, Figures 1.98 to 1.102) nearly the whole of the fore half corresponds to the head, and the hind half to the trunk.

The number of the metamera, and of the embryonic somites or primitive segments from which they develop, varies considerably in the vertebrates, according as the hind part of the body is short or is lengthened by a tail. In the developed man the trunk (including the rudimentary tail) consists of thirty–three metamera, the solid centre of which is formed by that number of vertebrae in the vertebral column (seven cervical, twelve dorsal, five lumbar, five sacral, and four caudal). To these we must add at least nine head–vertebrae, which originally (in all the craniota) constitute the skull. Thus the total number of the primitive segments of the human body is raised to at least forty–two; it would reach forty–five to forty–eight if (according to recent investigations) the number of the original segments of the skull is put at twelve to fifteen. In the tailless or anthropoid apes the number of metamera is much the same as in man, only differing by one or two; but it is much larger in the long–tailed apes and most of the other mammals. In long serpents and fishes it reaches several hundred (sometimes 400).

(FIGURES 1.158 TO 1.160. Embryo of the amphioxus, twenty four hours old, with eight somites. (From Hatschek.) Figures 1.158 and 1.159 lateral view (from left). Figure 1.160 seen from back. In Figure 1.158 only the outlines of the eight primitive segments are indicated, in Figure 1.159 their cavities and muscular walls. V fore end, H hind end, d gut, du under and dd upper wall of the gut, ne canalis neurentericus, nv ventral, nd dorsal wall of the neural tube, np neuroporus, dv fore pouch of the gut, ch chorda, mf mesodermic fold, pm polar cells of the mesoderm (ms), e ectoderm.)

In order to understand properly the real nature and origin of articulation in the human body and that of the higher vertebrates, it is necessary to compare it with that of the lower vertebrates, and bear in mind always the genetic connection of all the members of the stem. In this the simple development of the invaluable amphioxus once more furnishes the key to the complex and cenogenetically modified embryonic processes of the craniota. The articulation of the amphioxus begins at an early stage—earlier than in the craniotes. The two coelom–pouches have hardly grown out of the primitive gut (Figure 1.156 c) when the blind fore part of it (farthest away from the primitive mouth, u) begins to separate by a transverse fold (s): this is the first primitive segment. Immediately afterwards the hind part of the coelom–pouches begins to divide into a series of pieces by new transverse folds (Figure 1.157). The foremost of these primitive segments (us1) is the first and oldest; in Figures 1.124 and 1.157 there are already five formed. They separate so rapidly, one behind the other, that eight pairs are formed within twenty–four hours of the beginning of development, and seventeen pairs twenty–four hours later. The number increases as the embryo grows and extends backwards, and new cells are formed constantly (at the primitive mouth) from the two primitive mesodermic cells (Figures 1.159 to 1.160).

(FIGURES 1.161 AND 1.162. Transverse section of shark–embryos (through the region of the kidneys). (From Wijhe and Hertwig.) In Figure 1.162 the dorsal segment–cavities (h) are already separated from the body–cavity (lh), but they are connected a little earlier (Figure 1.161), nr neural tube, ch chorda, sch subchordal string, ao aorta, sk skeletal–plate, mp muscle–plate, cp cutis–plate, w connection of latter (growth–zone), vn primitive kidneys, ug prorenal duct, uk prorenal canals, us point where they are cut off, tr prorenal funnel, mk middle germ–layer (mk1 parietal, mk2 visceral), ik inner germ–layer (gut–gland layer).)

This typical articulation of the two coelom–sacs begins very early in the lancelet, before they are yet severed from the primitive gut, so that at first each segment–cavity (us) still communicates by a narrow opening with the gut, like an intestinal gland. But this opening soon closes by complete severance, proceeding regularly backwards. The closed segments then extend more, so that their upper half grows upwards like a fold between the ectoderm (ak) and neural tube (n), and the lower half between the ectoderm and alimentary canal (ch; Figure 1.82 d, left half of the figure). Afterwards the two halves completely separate, a lateral longitudinal fold cutting between them (mk, right half of Figure 1.82). The dorsal segments (sd) provide the muscles of the trunk the whole length of the body (1.159): this cavity afterwards disappears. On the other hand, the ventral parts give rise, from their uppermost section, to the pronephridia or primitive–kidney canals, and from the lower to the segmental rudiments of the sexual glands or gonads. The partitions of the muscular dorsal pieces (myotomes) remain, and determine the permanent articulation of the vertebrate organism. But the partitions of the large ventral pieces (gonotomes) become thinner, and afterwards disappear in part, so that their cavities run together to form the metacoel, or the simple permanent body–cavity.

The articulation proceeds in substantially the same way in the other vertebrates, the craniota, starting from the coelom–pouches. But whereas in the former case there is first a transverse division of the coelom–sacs (by vertical folds) and then the dorso–ventral division, the procedure is reversed in the craniota; in their case each of the long coelom–pouches first divides into a dorsal (primitive segment plates) and a ventral (lateral plates) section by a lateral longitudinal fold. Only the former are then broken up into primitive segments by the subsequent vertical folds; while the latter (segmented for a time in the amphioxus) remain undivided, and, by the divergence of their parietal and visceral plates, form a body–cavity that is unified from the first. In this case, again, it is clear that we must regard the features of the younger craniota as cenogenetically modified processes that can be traced palingenetically to the older acrania.

We have an interesting intermediate stage between the acrania and the fishes in these and many other respects in the cyclostoma (the hag and the lamprey, cf. Chapter 2.21).

(FIGURE 1.163. Frontal (or horizontal–longitudinal) section of a triton–embryo with three pairs of primitive segments. ch chorda, us primitive segments, ush their cavity, ak horn plate.)

Among the fishes the selachii, or primitive fishes, yield the most important information on these and many other phylogenetic questions (Figures 1.161 and 1.162). The careful studies of Ruckert, Van Wijhe, H.E. Ziegler, and others, have given us most valuable results. The products of the middle germinal layer are partly clear in these cases at the period when the dorsal primitive segment cavities (or myocoels, h) are still connected with the ventral body–cavity (lh; Figure 1.161). In Figure 1.162, a somewhat older embryo, these cavities are separated. The outer or lateral wall of the dorsal segment yields the cutis–plate (cp), the foundation of the connective corium. From its inner or median wall are developed the muscle–plate (mp, the rudiment of the trunk–muscles) and the skeletal plate, the formative matter of the vertebral column (sk).

In the amphibia, also, especially the water–salamander (Triton), we can observe very clearly the articulation of the coelom–pouches and the rise of the primitive segments from their dorsal half (cf. Figure 1.91, A, B, C). A horizontal longitudinal section of the salamander–embryo (Figure 1.163) shows very clearly the series of pairs of these vesicular dorsal segments, which have been cut off on each side from the ventral side–plates, and lie to the right and left of the chorda.

(FIGURE 1.164. The third cervical vertebra (human).

FIGURE 1.165. The sixth dorsal vertebra (human).

FIGURE 1.166. The second lumbar vertebra (human).)

The metamerism of the amniotes agrees in all essential points with that of the three lower classes of vertebrates we have considered; but it varies considerably in detail, in consequence of cenogenetic disturbances that are due in the first place (like the degeneration of the coelom–pouches) to the large development of the food–yelk. As the pressure of this seems to force the two middle layers together from the start, and as the solid structure of the mesoderm apparently belies the original hollow character of the sacs, the two sections of the mesoderm, which are at that time divided by the lateral fold—the dorsal segment–plates and ventral side–plates—have the appearance at first of solid layers of cells (Figures 1.94 to 1.97). And when the articulation of the somites begins in the sole–shaped embryonic shield, and a couple of protovertebrae are developed in succession, constantly increasing in number towards the rear, these cube–shaped somites (formerly called protovertebrae, or primitive vertebrae) have the appearance of solid dice, made up of mesodermic cells (Figure 1.93). Nevertheless, there is for a time a ventral cavity, or provertebral cavity, even in these solid "protovertebrae" (Figure 1.143 uwh). This vesicular condition of the provertebra is of the greatest phylogenetic interest; we must, according to the coelom theory, regard it as an hereditary reproduction of the hollow dorsal somites of the amphioxus (Figures 1.156 to 1.160) and the lower vertebrates (Figures 1.161 to 1.163). This rudimentary "provertebral cavity" has no physiological significance whatever in the amniote–embryo; it soon disappears, being filled up with cells of the muscular plate.

(FIGURE 1.167. Head of a shark embryo (Pristiurus), one–third of an inch long, magnified twenty times. (From Parker.) Seen from the ventral side.)

The innermost median part of the primitive segment plates, which lies immediately on the chorda (Figure 1.145 ch) and the medullary tube (m), forms the vertebral column in all the higher vertebrates (it is wanting in the lowest); hence it may be called the skeleton plate. In each of the provertebrae it is called the "sclerotome" (in opposition to the outlying muscular plate, the "myotome"). From the phylogenetic point of view the myotomes are much older than the sclerotomes. The lower or ventral part of each sclerotome (the inner and lower edge of the cube–shaped provertebra) divides into two plates, which grow round the chorda, and thus form the foundation of the body of the vertebra (wh). The upper plate presses between the chorda and the medullary tube, the lower between the chorda and the alimentary canal (Figure 1.137 C). As the plates of two opposite provertebral pieces unite from the right and left, a circular sheath is formed round this part of the chorda. From this develops the BODY of a vertebra—that is to say, the massive lower or ventral half of the bony ring, which is called the "vertebra" proper and surrounds the medullary tube (Figures 1.164 to 1.166). The upper or dorsal half of this bony ring, the vertebral arch (Figure 1.145 wb), arises in just the same way from the upper part of the skeletal plate, and therefore from the inner and upper edge of the cube–shaped primitive vertebra. As the upper edges of two opposing somites grow together over the medullary tube from right and left, the vertebra–arch becomes closed.

The whole of the secondary vertebra, which is thus formed from the union of the skeletal plates of two provertebral pieces and encloses a part of the chorda in its body, consists at first of a rather soft mass of cells; this afterwards passes into a firmer, cartilaginous stage, and finally into a third, permanent, bony stage. These three stages can generally be distinguished in the greater part of the skeleton of the higher vertebrates; at first most parts of the skeleton are soft, tender, and membranous; they then become cartilaginous in the course of their development, and finally bony.

(FIGURES 1.168 AND 1.169. Head of a chick embryo, of the third day. Figure 1.168 from the front, Figure 1.169 from the right. n rudimentary nose (olfactory pit), l rudimentary eye (optic pit, lens–cavity), g rudimentary ear (auditory pit), v fore–brain, gl

209

eye–cleft. Of the three pairs of gill–arches the first has passed into a process of the upper jaw (o) and of the lower jaw (u). (From Kolliker.))

At the head part of the embryo in the amniotes there is not generally a cleavage of the middle germinal layer into provertebral and lateral plates, but the dorsal and ventral somites are blended from the first, and form what are called the "head–plates" (Figure 1.148 k). From these are formed the skull, the bony case of the brain, and the muscles and corium of the body. The skull develops in the same way as the membranous vertebral column. The right and left halves of the head curve over the cerebral vesicle, enclose the foremost part of the chorda below, and thus finally form a simple, soft, membranous capsule about the brain. This is afterwards converted into a cartilaginous primitive skull, such as we find permanently in many of the fishes. Much later this cartilaginous skull becomes the permanent bony skull with its various parts. The bony skull in man and all the other amniotes is more highly differentiated and modified than that of the lower vertebrates, the amphibia and fishes. But as the one has arisen phylogenetically from the other, we must assume that in the former no less than the latter the skull was originally formed from the sclerotomes of a number of (at least nine) head–somites.

While the articulation of the vertebrate body is always obvious in the episoma or dorsal body, and is clearly expressed in the segmentation of the muscular plates and vertebrae, it is more latent in the hyposoma or ventral body. Nevertheless, the hyposomites of the vegetal half of the body are not less important than the episomites of the animal half. The segmentation in the ventral cavity affects the following principal systems of organs: 1, the gonads or sex–glands (gonotomes); 2, the nephridia or kidneys (nephrotomes); and 3, the head–gut with its gill–clefts (branchiotomes).

(FIGURE 1.170. Head of a dog embryo, seen from the front. a the two lateral halves of the foremost cerebral vesicle, b rudimentary eye, c middle cerebral vesicle, de first pair of gill–arches (e upper–jaw process, d lower–jaw process), f, f apostrophe, f double apostrophe, second, third, and fourth pairs of gill–arches, g h i k heart (g right, h left auricle; i left, k right ventricle), l origin of the aorta with three pairs of arches, which go to the gill–arches. (From Bischoff.))

The metamerism of the hyposoma is less conspicuous because in all the craniotes the cavities of the ventral segments, in the walls of which the sexual products are developed, have long since coalesced, and formed a single large body–cavity, owing to the disappearance of the partition. This cenogenetic process is so old that the cavity seems to be unsegmented from the first in all the craniotes, and the rudiment of the gonads also is almost always unsegmented. It is the more interesting to learn that, according to the important discovery of Ruckert, this sexual structure is at first segmental even in the actual selachii, and the several gonotomes only blend into a simple sexual gland on either side secondarily.

(FIGURE 1.171. Human embryo of the fourth week (twenty–six days old), one–fourth of an inch in length magnified twenty times, showing: point of development of the hind–leg, umbilical cord (underneath it the tail, bent upwards), trigeminal nerve V Trigeminus, optic–muscle nerve III Oculo–motorius, rolling muscle nerve IV Trochlearis, rudiment of ear (labyrinthic vesicles), pneumogastric nerve X Vagus, terminal nerve XI Accessorius, hypoglossal nerve XII Hypoglossus, first spinal nerve, point of development of arm (or fore–leg), true spinal nerve. (From Moll.) The rudiments of the cerebral nerves and the roots of the spinal nerves are especially marked. Underneath the four gill–arches (left side) is the heart (with auricle, V and ventricle, K), under this again the liver (L).)

Amphioxus, the sole surviving representative of the acrania, once more yields us most interesting information; in this case the sexual glands remain segmented throughout life. The sexually mature lancelet has, on the right and left of the gut, a series of metamerous sacs, which are filled with ova in the female and sperm in the male. These segmental gonads are originally nothing else than the real gonotomes, separate body–cavities, formed from the hyposomites of the trunk.

The gonads are the most important segmental organs of the hyposoma, in the sense that they are phylogenetically the oldest. We find sexual glands (as pouch–like appendages of the gastro–canal system) in most of the lower animals, even in the medusae, etc., which have no kidneys. The latter appear first (as a pair of excretory tubes) in the platodes (turbellaria), and have probably been inherited from these by the articulates (annelids) on the one hand and the unarticulated prochordonia on the other, and from these passed to the articulated vertebrates. The oldest form of the kidney system in this stem are the segmental pronephridia or prorenal canals, in the same arrangement as Boveri found them in the amphioxus. They are small canals that lie in the frontal plane, on each side of the chorda, between the episoma and hyposoma (Figure 1.102 n); their internal funnel–shaped opening leads into the various body–cavities, their outer opening is the lateral furrow of the epidermis. Originally they must have had a double function, the carrying away of the urine from the episomites and the release of the sexual cells from the hyposomites.

The recent investigations of Ruckert and Van Wijhe on the mesodermic segments of the trunk and the excretory system of the selachii show that these "primitive fishes" are closely related to the amphioxus in this further respect. The transverse section of the shark–embryo in Figure 1.161 shows this very clearly.

In other higher vertebrates, also, the kidneys develop (though very differently formed later on) from similar structures, which have been secondarily derived from the segmental pronephridia of the acrania. The parts of the mesoderm at which the first traces of them are found are usually called the middle or mesenteric plates. As the first traces of the gonads make their appearance in the lining of these middle plates nearer inward (or the middle) from the inner funnels of the nephro–canals, it is better to count this part of the mesoderm with the hyposoma.

The chief and oldest organ of the vertebrate hyposoma, the alimentary canal, is generally described as an unsegmented organ. But we could just as well say that it is the oldest of all the segmented organs of the vertebrate; the double row of the coelom–pouches grows out of the dorsal wall of the gut, on either side of the chorda. In the brief period during which these segmental coelom–pouches are still openly connected with the gut, they look just like a double chain of segmented visceral glands. But apart from this, we have originally in all vertebrates an important articulation of the fore–gut, that is wanting in the lower gut, the segmentation of the branchial (gill) gut.

(FIGURE 1.172. Transverse section of the shoulder and fore–limb (wing) of a chick–embryo of the fourth day, magnified about twenty times. Beside the medullary tube we can see on each side three clear streaks in the dark dorsal wall, which advance into the rudimentary fore–limb or wing (e). The uppermost of them is the muscular plate; the middle is the hind and the lowest the fore root of a spinal nerve. Under the chorda in the middle is the single aorta, at each side of it a cardinal vein, and below these the primitive kidneys. The gut is almost closed. The ventral wall advances into the amnion, which encloses the embryo. (From Remak.)

FIGURE 1.173. Transverse section of the pelvic region and hind legs of a chick–embryo of the fourth day, magnified about forty times. h horn–plate, w medullary tube, n canal of the tube, u primitive kidneys, x chorda, e hind legs, b allantoic canal in the ventral wall, t aorta, v cardinal veins, a gut, d gut–gland layer, f gut–fibre layer, g embryonic epithelium, r dorsal muscles, c body–cavity or coeloma. (From Waldeyer.))

The gill–clefts, which originally in the older acrania pierced the wall of the fore–gut, and the gill–arches that separated them, were presumably also segmental, and distributed among the various metamera of the chain, like the gonads in the after–gut and the nephridia. In the amphioxus, too, they are still segmentally formed. Probably there was a division of labour of the hyposomites in the older (and long extinct) acrania, in such wise that those of the fore–gut took over the function of breathing and those of the after–gut that of reproduction. The former developed into gill–pouches, the latter into sex–pouches. There may have been primitive kidneys in both. Though the gills have lost their function in the higher animals, certain parts of them have been generally maintained in the embryo by a tenacious heredity. At a very early stage we notice in the embryo of man and the other amniotes, at each side of the head, the remarkable and important structures which we call the gill–arches and gill–clefts (Figures 1.167 to 1.170 f). They belong to the characteristic and inalienable organs of the amniote–embryo, and are found always in the same spot and with the same arrangement and structure. There are formed to the right and left in the lateral wall of the fore–gut cavity, in its foremost part, first a pair and then several pairs of sac–shaped inlets, that pierce the whole thickness of the lateral wall of the head. They are thus converted into clefts, through which one can penetrate freely from without into the gullet. The wall thickens between these branchial folds, and changes into an arch–like or sickle–shaped piece—the gill, or gullet–arch. In this the muscles and skeletal parts of the branchial gut separate; a blood–vessel arch rises afterwards on their inner side (Figure 1.98 ka). The number of the branchial arches and the clefts that alternate with them is four or five on each side in the higher vertebrates (Figure 1.170 d, f, f apostrophe, f double apostrophe). In some of the fishes (selachii) and in the cyclostoma we find six or seven of them permanently.

These remarkable structures had originally the function of respiratory organs—gills. In the fishes the water that serves for breathing, and is taken in at the mouth, still always passes out by the branchial clefts at the sides of the gullet. In the higher vertebrates they afterwards disappear. The branchial arches are converted partly into the jaws, partly into the bones of the tongue and the ear. From the first gill–cleft is formed the tympanic cavity of the ear.

There are few parts of the vertebrate organism that, like the outer covering or integument of the body, are not subject to metamerism. The outer skin (epidermis) is unsegmented from the first, and proceeds from the continuous horny plate. Moreover, the underlying cutis is also not metamerous, although it develops from the segmental structure of the cutis–plates (Figures 1.161 and 1.162 cp). The vertebrates are strikingly and profoundly different from the articulates in these respects also.

Further, most of the vertebrates still have a number of unarticulated organs, which have arisen locally, by adaptation of particular parts of the body to certain special functions. Of this character are the sense–organs in the episoma, and the limbs, the heart, the spleen, and the large visceral glands—lungs, liver, pancreas, etc.—in the hyposoma. The heart is originally only a local spindle–shaped enlargement of the large ventral blood–vessel or principal vein, at the point where the subintestinal passes into the branchial artery, at the limit of the head and trunk (Figures 1.170 and 1.171). The three higher sense–organs—nose, eye, and ear—were originally developed in the same form in all the craniotes, as three pairs of small depressions in the skin at the side of the head.

The organ of smell, the nose, has the appearance of a pair of small pits above the mouth–aperture, in front of the head (Figure 1.169 n). The organ of sight, the eye, is found at the side of the head, also in the shape of a depression (Figures 1.169 l and 1.170 b), to which corresponds a large outgrowth of the foremost cerebral vesicle on each side. Farther behind, at each side of the head, there is a third depression, the first trace of the organ of hearing (Figure 1.169 g). As yet we can see nothing of the later elaborate structure of these organs, nor of the characteristic build of the face.

(FIGURE 1.174. Development of the lizard's legs (Lacerta agilis), with special relation to their blood–vessels. 1, 3, 5, 7, 9, 11 right fore–leg; 13, 15 left fore–leg; 2, 4, 6, 8, 10, 12 right hind–leg; 14, 16 left hind–leg; SRV lateral veins of the trunk, VU umbilical vein. (From F. Hochstetter.))

When the human embryo has reached this stage of development, it can still scarcely be distinguished from that of any other higher vertebrate. All the chief parts of the body are now laid down: the head with the primitive skull, the rudiments of the three higher sense–organs and the five cerebral vesicles, and the gill–arches and clefts; the trunk with the spinal cord, the rudiment of the vertebral column, the chain of metamera, the heart and chief blood–vessels, and the kidneys. At this stage man is a higher vertebrate, but shows no essential morphological difference from the embryos of the mammals, the birds, the reptiles, etc. This is an ontogenetic fact of the utmost significance. From it we can gather the most important phylogenetic conclusions.

There is still no trace of the limbs. Although head and trunk are separated and all the principal internal organs are laid down, there is no indication whatever of the "extremities" at this stage; they are formed later on. Here again we have a fact of the utmost interest. It proves that the older vertebrates had no feet, as we find to be the case in the lowest living vertebrates (amphioxus and the cyclostoma). The descendants of these ancient footless vertebrates only acquired extremities—two fore–legs and two hind–legs—at a much later stage of development. These were at first all alike, though they afterwards vary considerably in structure—becoming fins (of breast and belly) in the fishes, wings and legs in the birds, fore and hind legs in the creeping animals, arms and legs in the apes and man. All these parts develop from the same simple original structure, which forms secondarily from the trunk–wall (Figures 1.172 and 1.173). They have always the appearance of two pairs of small buds, which represent at first simple roundish knobs or plates. Gradually each of these plates becomes a large projection, in which we can distinguish a small inner part and a broader outer part. The latter is the rudiment of the foot or hand, the former that of the leg or arm. The similarity of the original rudiment of the limbs in different groups of vertebrates is very striking.

(FIGURE 1.175. Human embryo, five weeks old, half an inch long, seen from the right, magnified ten times. (From Russel Bardeen and Harmon Lewis.) In the undissected head we see the eye, mouth, and ear. In the trunk the skin and part of the muscles have been removed, so that the cartilaginous vertebral column is free; the dorsal root of a spinal nerve goes out from each vertebra (towards the skin of the back). In the middle of the lower half of the figure part of the ribs and intercostal muscles are visible. The skin and muscles have also been removed from the right limbs; the internal rudiments of the five fingers of the hand, and five toes of the foot, are clearly seen within the fin–shaped plate, and also the strong network of nerves that goes from the spinal cord to the extremities. The tail projects under the foot, and to the right of it is the first part of the umbilical cord.)

How the five fingers or toes with their blood–vessels gradually differentiate within the simple fin–like structure of the limbs can be seen in the instance of the lizard in Figure 1.174. They are formed in just the same way in man: in the human embryo of five weeks the five fingers can clearly be distinguished within the fin–plate (Figure 1.175).

The careful study and comparison of human embryos with those of other vertebrates at this stage of development is very instructive, and reveals more mysteries to the impartial student than all the religions in the world put together. For instance, if we compare attentively the three successive stages of development that are represented, in twenty different amniotes we find a remarkable likeness. When we see that as a fact twenty different amniotes of such divergent characters develop from the same embryonic form, we can easily understand that they may all descend from a common ancestor.

(FIGURES 1.176 TO 1.178. Embryos of the bat (Vespertilio murinus) at three different stages. (From Oscar Schultze.) Figure 1.176: Rudimentary limbs (v fore–leg, h hind–leg). l lenticular depression, r olfactory pit, ok upper jaw, uk lower jaw, k2, k3, k4 first, second and third gill–arches, a amnion, n umbilical vessel, d yelk–sac. Figure 1.177: Rudiment of flying membrane, membranous fold between fore and hind leg. n umbilical vessel, o ear–opening, f flying membrane. Figure 1.178: The flying membrane developed and stretched across the fingers of the hands, which cover the face.)

In the first stage of development, in which the head with the five cerebral vesicles is already clearly indicated, but there are no limbs, the embryos of all the vertebrates, from the fish to man, are only incidentally or not at all different from each other. In the second stage, which shows the limbs, we begin to see differences between the embryos of the lower and higher vertebrates; but the human embryo is still hardly distinguishable from that of the higher mammals. In the third stage, in which the gill–arches have disappeared and the face is formed, the differences become more pronounced. These are facts of a significance that cannot be exaggerated.* (* Because they show how the most diverse structures may be developed from a common form. As we actually see this in the case of the embryos, we have a right to assume it in that of the stem–forms. Nevertheless, this resemblance, however great, is never a real identity. Even the embryos of the different individuals of one species are usually not really identical. If the reader can consult the complete edition of this work at a library, he will find six plates illustrating these twenty embryos.)

If there is an intimate causal connection between the processes of embryology and stem–history, as we must assume in virtue of the laws of heredity, several important phylogenetic conclusions follow at once from these ontogenetic facts. The profound and remarkable similarity in the embryonic development of man and the other vertebrates can only be explained when we admit their descent from a common ancestor. As a fact, this common descent is now accepted by all competent scientists; they have substituted the natural evolution for the supernatural creation of organisms.

CHAPTER 1.15. FOETAL MEMBRANES AND CIRCULATION.

Among the many interesting phenomena that we have encountered in the course of human embryology, there is an especial importance in the fact that the development of the human body follows from the beginning just the same lines as that of the other viviparous mammals. As a fact, all the embryonic peculiarities that distinguish the mammals from other animals are found also in man; even the ovum with its distinctive membrane (zona pellucida, Figure 1.14) shows the same typical structure in all mammals (apart from the older oviparous monotremes). It has long since been deduced from the structure of the developed man that his natural place in the animal kingdom is among the mammals. Linne (1735) placed him in this class with the apes, in one and the same order (primates), in his Systema Naturae. This position is fully confirmed by comparative embryology. We see that man entirely resembles the higher mammals, and most of all the apes, in embryonic development as well as in anatomic structure. And if we seek to understand this ontogenetic agreement in the light of the biogenetic law, we find that it proves clearly and necessarily the descent of man from a series of other mammals, and proximately from the primates. The common origin of man and the other mammals from a single ancient stem–form can no longer be questioned; nor can the immediate blood–relationship of man and the ape.

(FIGURE 1.179. Human embryos from the second to the fifteenth week, natural size, seen from the left, the curved back turned towards the right. (Mostly from Ecker.) II of fourteen days. III of three weeks. IV of four weeks. V of five weeks. VI of six weeks. VII of seven weeks. VIII of eight weeks. XII of twelve weeks. XV of fifteen weeks.)

The essential agreement in the whole bodily form and inner structure is still visible in the embryo of man and the other mammals at the late stage of development at which the mammal–body can be recognised as such. But at a somewhat earlier stage, in which the limbs, gill–arches, sense–organs, etc., are already outlined, we cannot yet recognise the mammal embryos as such, or distinguish them from those of birds and reptiles. When we consider still earlier stages of development, we are unable to discover any essential difference in bodily structure between the embryos of these higher vertebrates and those of the lower, the amphibia and fishes. If, in fine, we go back to the construction of the body out of the four germinal layers, we are astonished to perceive that these four layers are the same in all vertebrates, and everywhere take a similar part in the building–up of the fundamental organs of the body. If we inquire as to the origin of these four secondary layers, we learn that they always arise in the same way from the two primary layers; and the latter have the same significance in all the metazoa (i.e., all animals except the unicellulars). Finally, we see that the cells which make up the primary germinal layers owe their origin in every case to the repeated cleavage of a single simple cell, the stem–cell or fertilised ovum.

(FIGURE 1.180. Very young human embryo of the fourth week, one–fourth of an inch long (taken from the womb of a suicide eight hours after death). (From Rabl.) n nasal pits, a eye, u lower jaw, z arch of hyoid bone, k3 and k4 third and fourth gill–arch, h heart; s primitive segments, vg fore–limb (arm), hg hind–limb (leg), between the two the ventral pedicle.)

It is impossible to lay too much stress on this remarkable agreement in the chief embryonic features in man and the other animals. We shall make use of it later on for our monophyletic theory of descent—the hypothesis of a common descent of man and all the metazoa from the gastraea. The first rudiments of the principal parts of the body, especially the oldest organ, the alimentary canal, are the same everywhere; they have always the same extremely simple form. All the peculiarities that distinguish the various groups of animals from each other only appear gradually in the course of embryonic development; and the closer the relation of the various groups, the later they are found. We may formulate this phenomenon in a definite law, which may in a sense be regarded as an appendix to our biogenetic law. This is the law of the ontogenetic connection of related animal forms. It runs: The closer the relation of two fully–developed animals in respect of their whole bodily structure, and the nearer they are connected in the classification of the animal kingdom, the longer do their embryonic forms retain their identity, and the longer is it impossible (or only possible on the ground of subordinate features) to distinguish between their embryos. This law applies to all animals whose embryonic development is, in the main, an hereditary summary of their ancestral history, or in which the original form of development has been faithfully preserved by heredity.

When, on the other hand, it has been altered by cenogenesis, or disturbance of development, we find a limitation of the law, which increases in proportion to the introduction of new features by adaptation (cf. Chapter 1.1). Thus the apparent exceptions to the law can always be traced to cenogenesis.

(FIGURE 1.181. Human embryo of the middle of the fifth week, one–third of an inch long. (From Rabl.) Letters as in Figure 1.180, except sk curve of skull, ok upper jaw, hb neck–indentation.)

When we apply to man this law of the ontogenetic connection of related forms, and run rapidly over the earliest stages of human development with an eye to it, we notice first of all the structural identity of the ovum in man and the other mammals at the very beginning (Figures 1.1 and 1.14). The human ovum possesses all the distinctive features of the ovum of the viviparous mammals, especially the characteristic formation of its membrane (zona pellucida), which clearly distinguishes it from the ovum of all other animals. When the human foetus has attained the age of fourteen days, it forms a round vesicle (or "embryonic vesicle") about a quarter of an inch in diameter. A thicker part of its border forms a simple sole–shaped embryonic shield one–twelfth of an inch long (Figure 1.133). On its dorsal side we find in the middle line the straight medullary furrow, bordered by the two parallel dorsal or medullary swellings. Behind, it passes by the neurenteric canal into the primitive gut or primitive groove. From this the folding of the two coelom–pouches proceeds in the same way as in the other mammals (cf. Figures 1.96 and 1.97). In the middle of the sole–shaped embryonic shield the first primitive segments immediately begin to make their appearance. At this age the human embryo cannot be distinguished from that of other mammals, such as the hare or dog.

A week later (or after the twenty–first day) the human embryo has doubled its length; it is now about one–fifth of an inch long, and, when seen from the side, shows the characteristic bend of the back, the swelling of the head–end, the first outline of the three higher sense–organs, and the rudiments of the gill–clefts, which pierce the sides of the neck (Figure 1.179, III). The allantois has grown out of the gut behind. The embryo is already entirely enclosed in the amnion, and is only connected in the middle of the belly by the vitelline duct with the embryonic vesicle, which changes into the yelk–sac. There are no extremities or limbs at this stage, no trace of arms or legs. The head–end has been strongly differentiated from the tail–end; and the first outlines of the cerebral vesicles in front, and the heart below, under the fore–arm, are already more or less clearly seen.

There is as yet no real face. Moreover, we seek in vain at this stage a special character that may distinguish the human embryo from that of other mammals.

(FIGURE 1.182. Median longitudinal section of the tail of a human embryo, two–thirds of an inch long. (From Ross Granville Harrison.) Med medullary tube, Ca.fil caudal filament, ch chorda, ao caudal artery, V.c.i caudal vein, an anus, S.ug sinus urogenitalis.)

A week later (after the fourth week, on the twenty–eighth to thirtieth day of development) the human embryo has reached a length of about one–third of an inch (Figure 1.179 IV). We can now clearly distinguish the head with its various parts; inside it the five primitive cerebral vesicles (fore–brain, middle–brain, intermediate–brain, hind–brain, and after–brain); under the head the gill–arches, which divide the gill–clefts; at the sides of the head the rudiments of the eyes, a couple of pits in the outer skin, with a pair of corresponding simple vesicles growing out of the lateral wall of the fore–brain (Figures 1.180, 1.181 a). Far behind the eyes, over the last gill–arches, we see a vesicular rudiment of the auscultory organ. The rudimentary limbs are now clearly outlined—four simple buds of the shape of round plates, a pair of fore (vg) and a pair of hind legs (hg), the former a little larger than the latter. The large head bends over the trunk, almost at a right angle. The latter is still connected in the middle of its ventral side with the embryonic vesicle; but the embryo has still further severed itself from it, so that it already hangs out as the yelk–sac. The hind part of the body is also very much curved, so that the pointed tail–end is directed towards the head. The head and face–part are sunk entirely on the still open breast. The bend soon increases so much that the tail almost touches the forehead (Figure 1.179 V.; Figure 1.181). We may then distinguish three or four special curves on the round dorsal surface—namely, a skull–curve in the region of the second cerebral vesicle, a neck–curve at the beginning of the spinal cord, and a tail–curve at the fore–end. This pronounced curve is only shared by man and the higher classes of vertebrates (the amniotes); it is much slighter, or not found at all, in the lower vertebrates. At this age (four weeks) man has a considerable tail, twice as long as his legs. A vertical longitudinal section through the middle plane of this tail (Figure 1.182) shows that the hinder end of the spinal marrow extends to the point of the tail, as also does the underlying chorda (ch), the terminal continuation of the vertebral column. Of the latter, the rudiments of the seven coccygeal (or lowest) vertebrae are visible—thirty–two indicates the third and thirty–six the seventh of these. Under the vertebral column we see the hindmost ends of the two large blood–vessels of the tail, the principal artery (aorta caudalis or arteria sacralis media, Ao), and the principal vein (vena caudalis or sacralis media). Underneath is the opening of the anus (an) and the urogenital sinus (S.ug). From this anatomic structure of the human tail it is perfectly clear that it is the rudiment of an ape–tail, the last hereditary relic of a long hairy tail, which has been handed down from our tertiary primate ancestors to the present day.

(FIGURE 1.183. Human embryo, four weeks old, opened on the ventral side. Ventral and dorsal walls are cut away, so as to show the contents of the pectoral and abdominal cavities. All the appendages are also removed (amnion, allantois, yelk–sac), and the middle part of the gut. n eye, 3 nose, 4 upper jaw, 5 lower jaw, 6 second, 6 double apostrophe, third gill–arch, ov heart (o right, o apostrophe, left auricle; v right, v apostrophe, left ventricle), b origin of the aorta, f liver (u umbilical vein), e gut (with vitelline artery, cut off at a apostrophe), j apostrophe, vitelline vein, m primitive kidneys, t rudimentary sexual glands, r terminal gut (cut off at the mesentery z), n umbilical artery, u umbilical vein, 9 fore–leg, 9 apostrophe, hind–leg. (From Coste.)

FIGURE 1.184. Human embryo, five weeks old, opened from the ventral side (as in Figure 1.183). Breast and belly–wall and liver are removed. 3 outer nasal process, 4 upper jaw, 5 lower jaw, z tongue, v right, v apostrophe, left ventricle of heart, o apostrophe, left auricle, b origin of aorta, b apostrophe, b double apostrophe, b triple apostrophe, first, second, and third aorta–arches, c, c apostrophe, c double apostrophe, vena cava, ae lungs (y pulmonary artery), e stomach, m primitive kidneys (j left vitelline vein, s cystic vein, a right vitelline artery, n umbilical artery, u umbilical vein), x vitelline duct, i rectum, 8 tail, 9 fore–leg, 9 apostrophe, hind–leg. (From Coste.))

It sometimes happens that we find even external relics of this tail growing. According to the illustrated works of Surgeon–General Bernhard Ornstein, of Greece, these tailed men are not uncommon; it is not impossible that they gave rise to the ancient fables of the satyrs. A great number of such cases are given by Max Bartels in his essay on "Tailed Men" (1884, in the Archiv fur Anthropologie, Band 15), and critically examined. These atavistic human tails are often mobile; sometimes they contain only muscles and fat, sometimes also rudiments of caudal vertebrae. They have a length of eight to ten inches and more. Granville Harrison has very carefully studied one of these cases of "pigtail," which he removed by operation from a six months old child in 1901. The tail moved briskly when the child cried or was excited, and was drawn up when at rest.

(FIGURE 1.185. The head of Miss Julia Pastrana. (From a photograph by Hintze.)

FIGURE 1.186. Human ovum of twelve to thirteen days (?). (From Allen Thomson.) 1. Not opened, natural size. 2. Opened and magnified. Within the outer chorion the tiny curved foetus lies on the large embryonic vesicle, to the left above.

FIGURE 1.187. Human ovum of ten days. (From Allen Thomson.) Natural size, opened; the small foetus in the right half, above.

FIGURE 1.188. Human foetus of ten days, taken from the preceding ovum, magnified ten times, a yelk–sac, b neck (the medullary groove already closed), c head (with open medullary groove), d hind part (with open medullary groove), e a shred of the amnion.

FIGURE 1.189. Human ovum of twenty to twenty–two days. (From Allen Thomson.) Natural size, opened. The chorion forms a spacious vesicle, to the inner wall of which the small foetus (to the right above) is attached by a short umbilical cord.

FIGURE 1.190. Human foetus of twenty to twenty–two days, taken from the preceding ovum, magnified. a amnion, b yelk–sac, c lower–jaw process of the first gill–arch, d upper–jaw process of same, e second gill–arch (two smaller ones behind). Three gill–clefts are clearly seen. f rudimentary fore–leg, g auditory vesicle, h eye, i heart.)

In the opinion of some travellers and anthropologists, the atavistic tail–formation is hereditary in certain isolated tribes (especially in south–eastern Asia and the archipelago), so that we might speak of a special race or "species" of tailed men (Homo caudatus). Bartels has "no doubt that these tailed men will be discovered in the advance of our geographical and ethnographical knowledge of the lands in question" (Archiv fur Anthropologie, Band 15 page 129).

When we open a human embryo of one month (Figure 1.183), we find the alimentary canal formed in the body–cavity, and for the most part cut off from the embryonic vesicle. There are both mouth and anus apertures. But the mouth–cavity is not yet separated from the nasal cavity, and the face not yet shaped. The heart shows all its four sections; it is very large, and almost fills the whole of the pectoral cavity (Figure 1.183 ov). Behind it are the very small rudimentary lungs. The primitive kidneys (m) are very large; they fill the greater part of the abdominal cavity, and extend from the liver (f) to the pelvic gut. Thus at the end of the first month all the chief organs are already outlined. But there are at this stage no features by which the human embryo materially differs from that of the dog, the hare, the ox, or the horse—in a word, of any other higher mammal. All these embryos have the same, or at least a very similar, form; they can at the most be distinguished from the human embryo by the total size of the body or some other insignificant difference in size. Thus, for instance, in man the head is larger in proportion to the trunk than in the ox. The tail is rather longer in the dog than in man. These are all negligible differences. On the other hand, the whole internal organisation and the form and arrangement of the various organs are essentially the same in the human embryo of four weeks as in the embryos of the other mammals at corresponding stages.

(FIGURE 1.191. Human embryo of sixteen to eighteen days. (From Coste.) Magnified. The embryo is surrounded by the amnion, (a), and lies free with this in the opened embryonic vesicle. The belly is drawn up by the large yelk–sac (d), and fastened to the inner wall of the embryonic membrane by the short and thick pedicle (b). Hence the normal convex curve of the back (Figure 1.190) is here changed into an abnormal concave surface. h heart, m parietal mesoderm. The spots on the outer wall of the serolemma are the roots of the branching chorion–villi, which are free at the border.

FIGURE 1.192. Human embryo of the fourth week, one–third of an inch long, lying in the dissected chorion.

FIGURE 1.193. Human embryo of the fourth week, with its membranes, like Figure 1.192, but a little older. The yelk–sac is rather smaller, the amnion and chorion larger.)

It is otherwise in the second month of human development. Figure 1.179 represents a human embryo of six weeks (VI), one of seven weeks (VII), and one of eight weeks (VIII), at natural size. The differences which mark off the human embryo from that of the dog and the lower mammals now begin to be more pronounced. We can see important differences at the sixth, and still more at the eighth week, especially in the formation of the head. The size of the various sections of the brain is greater in man, and the tail is shorter. Other differences between man and the lower mammals are found in the relative size of the internal organs. But even at this stage the human embryo differs very little from that of the nearest related mammals—the apes, especially the anthropomorphic apes. The features by means of which we distinguish between them are not clear until later on. Even at a much more advanced stage of development, when we can distinguish the human foetus from that of the ungulates at a glance, it still closely resembles that of the higher apes. At last we get the distinctive features, and we can distinguish the human

223

embryo confidently at the first glance from that of all other mammals during the last four months of foetal life—from the sixth to the ninth month of pregnancy. Then we begin to find also the differences between the various races of men, especially in regard to the formation of the skull and the face. (Cf. Chapter 2.23.)

(FIGURE 1.194. Human embryo with its membranes, six weeks old. The outer envelope of the whole ovum is the chorion, thickly covered with its branching villi, a product of the serous membrane. The embryo is enclosed in the delicate amnion—sac. The yelk—sac is reduced to a small pear—shaped umbilical vesicle; its thin pedicle, the long vitelline duct, is enclosed in the umbilical cord. In the latter, behind the vitelline duct, is the much shorter pedicle of the allantois, the inner lamina of which (the gut—gland layer) forms a large vesicle in most of the mammals, while the outer lamina is attached to the inner wall of the outer embryonic coat, and forms the placenta there. (Half diagrammatic.))

The striking resemblance that persists so long between the embryo of man and of the higher apes disappears much earlier in the lower apes. It naturally remains longest in the large anthropomorphic apes (gorilla, chimpanzee, orang, and gibbon). The physiognomic similarity of these animals, which we find so great in their earlier years, lessens with the increase of age. On the other hand, it remains throughout life in the remarkable long—nosed ape of Borneo (Nasalis larvatus). Its finely—shaped nose would be regarded with envy by many a man who has too little of that organ. If we compare the face of the long—nosed ape with that of abnormally ape—like human beings (such as the famous Miss Julia Pastrana, Figure 1.185), it will be admitted to represent a higher stage of development. There are still people among us who look especially to the face for the "image of God in man." The long—nosed ape would have more claim to this than some of the stumpy—nosed human individuals one meets.

This progressive divergence of the human from the animal form, which is based on the law of the ontogenetic connection between related forms, is found in the structure of the internal organs as well as in external form. It is also expressed in the construction of the envelopes and appendages that we find surrounding the foetus externally, and that we will now consider more closely. Two of these appendages—the amnion and the allantois—are only found in the three higher classes of vertebrates, while the third, the yelk—sac, is found in most of the vertebrates. This is a circumstance of great importance, and it gives us valuable data for constructing man's genealogical tree.

(FIGURE 1.195. Diagram of the embryonic organs of the mammal (foetal membranes and appendages). (From Turner.) E, M, H outer, middle, and inner germ layer of the embryonic shield, which is figured in median longitudinal section, seen from the left. am amnion. AC amniotic cavity, UV yelk–sac or umbilical vesicle, ALC allantois, al pericoelom or serocoelom (inter–amniotic cavity), sz serolemma (or serous membrane), pc prochorion (with villi).)

As regards the external membrane that encloses the ovum in the mammal womb, we find it just the same in man as in the higher mammals. The ovum is, the reader will remember, first surrounded by the transparent structureless ovolemma or zona pellucida (Figures 1.1 and 1.14). But very soon, even in the first week of development, this is replaced by the permanent chorion. This is formed from the external layer of the amnion, the serolemma, or "serous membrane," the formation of which we shall consider presently; it surrounds the foetus and its appendages as a broad, completely closed sac; the space between the two, filled with clear watery fluid, is the serocoelom, or interamniotic cavity ("extra–embryonic body–cavity"). But the smooth surface of the sac is quickly covered with numbers of tiny tufts, which are really hollow outgrowths like the fingers of a glove (Figures 1.186, 1.191 and 1.198 chz). They ramify and push into the corresponding depressions that are formed by the tubular glands of the mucous membrane of the maternal womb. Thus, the ovum secures its permanent seat (Figures 1.186 to 1.194).

In human ova of eight to twelve days this external membrane, the chorion, is already covered with small tufts or villi, and forms a ball or spheroid of one–fourth to one–third of an inch in diameter (Figures 1.186 to 1.188). As a large quantity of fluid gathers inside it, the chorion expands more and more, so that the embryo only occupies a small part of the space within the vesicle. The villi of the chorion grow larger and more numerous. They branch out more and more. At first the villi cover the whole surface, but they afterwards disappear from the greater part of it; they then develop with proportionately greater vigour at a spot where the placenta is formed from the allantois.

When we open the chorion of a human embryo of three weeks, we find on the ventral side of the foetus a large round sac, filled with fluid. This is the yelk–sac, or "umbilical vesicle," the origin of which we have considered previously. The larger the embryo becomes the smaller we find the yelk–sac. In the end we find the remainder of it in the shape of a small pear–shaped vesicle, fastened to a long thin stalk (or pedicle), and hanging from the open belly of the foetus (Figure 1.194). This pedicle is the vitelline duct, and is separated from the body at the closing of the navel.

Behind the yelk–sac a second appendage, of much greater importance, is formed at an early stage at the belly of the mammal embryo. This is the allantois or "primitive urinary sac," an important embryonic organ, only found in the three higher classes of vertebrates. In all the amniotes the allantois quickly appears at the hinder end of the alimentary canal, growing out of the cavity of the pelvic gut (Figure 1.147 r, u, Figure 1.195 ALC}.

(FIGURE 1.196. Diagrammatic frontal section of the pregnant human womb. (From Longet.) The embryo hangs by the umbilical cord, which encloses the pedicle of the allantois (al). nb umbilical vessel, am amnion, ch chorion, ds decidua serotina, dv decidua vera, dr decidua reflexa, z villi of the placenta, c cervix uteri, u uterus.)

The further development of the allantois varies considerably in the three sub–classes of the mammals. The two lower sub–classes, monotremes and marsupials, retain the simpler structure of their ancestors, the reptiles. The wall of the allantois and the enveloping serolemma remains smooth and without villi, as in the birds. But in the third sub–class of the mammals the serolemma forms, by invagination at its outer surface, a number of hollow tufts or villi, from which it takes the name of the chorion or mallochorion. The gut–fibre layer of the allantois, richly supplied with branches of the umbilical vessel, presses into these tufts of the primary chorion, and forms the "secondary chorion." Its embryonic blood–vessels are closely correlated to the contiguous maternal blood–vessels of the environing womb, and thus is formed the important nutritive apparatus of the embryo which we call the placenta.

The pedicle of the allantois, which connects the embryo with the placenta and conducts the strong umbilical vessels from the former to the latter, is covered by the amnion, and, with this amniotic sheath and the pedicle of the yelk–sac, forms what is called the umbilical cord (Figure 1.196 al). As the large and blood–filled vascular network of the foetal allantois attaches itself closely to the mucous lining of the maternal womb, and the partition between the blood–vessels of mother and child becomes much thinner, we get that remarkable nutritive apparatus of the foetal body which is characteristic of the placentalia (or choriata). We shall return afterwards to the closer consideration of this (cf. Chapter 2.23).

In the various orders of mammals the placenta undergoes many modifications, and these are in part of great evolutionary importance and useful in classification. There is only one of these that need be specially mentioned—the important fact, established by Selenka in 1890, that the distinctive human placentation is confined to the anthropoids. In this most advanced group of the mammals the allantois is very small, soon loses its cavity, and then, in common with the amnion, undergoes certain peculiar changes. The umbilical cord develops in this case from what is called the "ventral pedicle." Until very recently this was regarded as a structure peculiar to man. We now know from Selenka that the much–discussed ventral pedicle is merely the pedicle of the allantois, combined with the pedicle of the amnion and the rudimentary pedicle of the yelk–sac. It has just the same structure in the orang and gibbon (Figure 1.197) and very probably in the chimpanzee and gorilla, as in man; it is, therefore, not a DISPROOF, but a striking fresh proof, of the blood–relationship of man and the anthropoid apes.

(FIGURE 1.197. Male embryo of the Siamang–gibbon (Hylobates siamanga) of Sumatra, two–thirds natural size; to the left the dissected uterus, of which only the dorsal half is given. The embryo has been taken out, and the limbs folded together; it is still connected by the umbilical cord with the centre of the circular placenta which is attached to the inside of the womb. This embryo takes the head–position in the womb, and this is normal in man also.)

We find only in the anthropoid apes—the gibbon and orang of Asia and the chimpanzee and gorilla of Africa—the peculiar and elaborate formation of the placenta that characterises man (Figure 1.198). In this case there is at an early stage an intimate blending of the chorion of the embryo and the part of the mucous lining of the womb to which it attaches. The villi of the chorion with the blood–vessels they contain grow so completely into the tissue of the uterus, which is rich in blood, that it becomes impossible to separate them, and they form together a sort of cake. This comes away as the "afterbirth" at parturition; at the same time, the part of the mucous lining of the womb that has united inseparably with the chorion is torn away; hence it is called the decidua ("falling–away membrane"), and also the "sieve–membrane," because it is perforated like a sieve. We find a decidua of this kind in most of the higher placentals; but it is only in man and the anthropoid apes that it divides into three parts—the outer, inner, and placental decidua. The external or true decidua (Figure 1.196 du, Figure 1.199 g) is the part of the mucous lining of the womb that clothes the inner surface of the uterine cavity wherever it is not connected with the placenta. The placental or spongy decidua (placentalis or serotina, Figure 1.196 ds, Figure 1.199 d) is really the placenta itself, or the maternal part of it (placenta uterina)—namely, that part of the mucous lining of the womb which unites intimately with the chorion–villi of the foetal placenta. The internal or false decidua (interna or reflexa, Figure 1.196 dr, Figure 1.199 f) is that part of the mucous lining of the womb which encloses the remaining surface of the ovum, the smooth chorion (chorion laeve), in the shape of a special thin membrane. The origin of these three different deciduous membranes, in regard to which quite erroneous views (still retained in their names) formerly prevailed, is now quite clear, The external decidua vera is the specially modified and subsequently detachable superficial stratum of the original mucous lining of the womb. The placental decidua serotina is that part of the preceding which is completely transformed by the ingrowth of the chorion–villi, and is used for constructing the placenta. The inner decidua reflexa is formed by the rise of a circular fold of the mucous lining (at the border of the decidua vera and serotina), which grows over the foetus (like the anmnion) to the end.

The peculiar anatomic features that characterise the human foetal membranes are found in just the same way in the higher apes. Until recently it was thought that the human embryo was distinguished by its peculiar construction of a solid allantois and a special ventral pedicle, and that the umbilical cord developed from this in a different way than in the other mammals. The opponents of the unwelcome "ape–theory" laid great stress on this, and thought they had at last discovered an important indication that separated man from all the other placentals. But the remarkable discoveries published by the distinguished zoologist Selenka in 1890 proved that man shares these peculiarities of placentation with the anthropoid apes, though they are not found in the other apes. Thus the very feature which was advanced by our critics as a disproof became a most important piece of evidence in favour of our pithecoid origin.)

(FIGURE 1.198. Frontal section of the pregnant human womb, showing: end of the decidua, uterine cavity, chorion (laeve), amniotic cavity, foetal placenta, oviduct, spongy decidua serotina, umbilical vesicle, amnion, decidua reflexa, decidua vera, muscular wall of the uterus, mouth of the uterus. (From Turner.) The embryo (a month old) hangs in the middle of the amniotic cavity by the ventral pedicle or umbilical cord, which connects it with the placenta (above).

FIGURE 1.199. Human foetus, twelve weeks old, with its membranes. Natural size. The umbilical cord goes from its navel to the placenta. b amnion, c chorion, d placenta, d apostrophe, relics of villi on smooth chorion, f internal or reflex decidua, g external or true decidua. (From B. Schultze.)

FIGURE 1.200. Mature human foetus (at the end of pregnancy, in its natural position, taken out of the uterine cavity). On the inner surface of the latter (to the left) is the placenta, which is connected by the umbilical cord with the child's navel. (From Bernhard Schultze.))

Of the three vesicular appendages of the amniote embryo which we have now described the amnion has no blood–vessels at any moment of its existence. But the other two vesicles, the yelk–sac and the allantois, are equipped with large blood–vessels, and these effect the nourishment of the embryonic body. We may take the opportunity to make a few general observations on the first circulation in the embryo and its central organ, the heart. The first blood–vessels, the heart, and the first blood itself, are formed from the gut–fibre layer. Hence it was called by earlier embryologists the "vascular layer." In a sense the term is quite correct. But it must not be understood as if all the blood–vessels in the body came from this layer, or as if the whole of this layer were taken up only with the formation of blood–vessels. Neither of these suppositions is true. Blood–vessels may be formed independently in other parts, especially in the various products of the skin–fibre layer.

The first blood–vessels of the mammal embryo have been considered by us previously, and we shall study the development of the heart in the second volume.

(FIGURE 1.201. Vitelline vessels in the germinative area of a chick–embryo, at the close of the third day of incubation. (From Balfour.) The detached germinative area is seen from the ventral side: the arteries are dark, the veins light. H heart, AA aorta–arches, Ao aorta, R.Of.A right omphalo–mesenteric artery, S.T sinus terminalis, L.Of and R.Of right and left omphalo–mesenteric veins, S.V sinus venosus, D.C ductus Cuvieri, S.Ca.V and V.Ca fore and hind cardinal veins.)

In every vertebrate it lies at first in the ventral wall of the fore–gut, or in the ventral (or cardiac) mesentery, by which it is connected for a time with the wall of the body. But it soon severs itself from the place of its origin, and lies freely in a cavity—the cardiac cavity. For a short time it is still connected with the former by the thin plate of the mesocardium. Afterwards it lies quite free in the cardiac cavity, and is only directly connected with the gut–wall by the vessels which issue from it.

The fore–end of the spindle–shaped tube, which soon bends into an S–shape (Figure 1.202), divides into a right and left branch. These tubes are bent upwards arch–wise, and represent the first arches of the aorta. They rise in the wall of the fore–gut, which they enclose in a sense, and then unite above, in the upper wall of the fore gut–cavity, to form a large single artery, that runs backward immediately under the chorda, and is called the aorta (Figure 1.201 Ao). The first pair of aorta–arches rise on the inner wall of the first pair of gill–arches, and so lie between the first gill–arch (k) and the fore–gut (d), just as we find them throughout life in the fishes. The single aorta, which results from the conjunction of these two first vascular arches, divides again immediately into two parallel branches, which run backwards on either side of the chorda. These are the primitive aortas which we have already mentioned; they are also called the posterior vertebral arteries. These two arteries now give off at each side, behind, at right angles, four or five branches, and these pass from the embryonic body to the germinative area, they are called omphalo–mesenteric or vitelline arteries. They represent the first beginning of a foetal circulation. Thus, the first blood–vessels pass over the embryonic body and reach as far as the edge of the germinative area. At first they are confined to the dark or "vascular" area. But they afterwards extend over the whole surface of the embryonic vesicle. In the end, the whole of the yelk–sac is covered with a vascular net–work. These vessels have to gather food from the contents of the yelk–sac and convey it to the embryonic body. This is done by the veins, which pass first from the germinative area, and afterwards from the yelk–sac, to the farther end of the heart. They are called vitelline, or, frequently, omphalo–mesenteric, veins.

These vessels naturally atrophy with the degeneration of the umbilical vesicle, and the vitelline circulation is replaced by a second, that of the allantois. Large blood–vessels are developed in the wall of the urinary sac or the allantois, as before, from the gut–fibre layer. These vessels grow larger and larger, and are very closely connected with the vessels that develop in the body of the embryo itself. Thus, the secondary, allantoic circulation gradually takes the place of the original vitelline circulation. When the allantois has attached itself to the inner wall of the chorion and been converted into the placenta, its blood–vessels alone effect the nourishment of the embryo. They are called umbilical vessels, and are originally double—a pair of umbilical arteries and a pair of umbilical veins. The two umbilical veins (Figure 1.183 u), which convey blood from the placenta to the heart, open it first into the united vitelline veins. The latter then disappear,

230

and the right umbilical vein goes with them, so that henceforth a single large vein, the left umbilical vein, conducts all the blood from the placenta to the heart of the embryo. The two arteries of the allantois, or the umbilical arteries (Figures 1.183 n and 1.184 n), are merely the ultimate terminations of the primitive aortas, which are strongly developed afterwards. This umbilical circulation is retained until the nine months of embryonic life are over, and the human embryo enters into the world as the independent individual. The umbilical cord (Figure 1.196 al), in which these large blood–vessels pass from the embryo to the placenta, comes away, together with the latter, in the after–birth, and with the use of the lungs begins an entirely new form of circulation, which is confined to the body of the infant.

(FIGURE 1.202. Boat–shaped embryo of the dog, from the ventral side, magnified about ten times. In front under the forehead we can see the first pair of gill–arches; underneath is the S–shaped heart, at the sides of which are the auditory vesicles. The heart divides behind into the two vitelline veins, which expand in the germinative area (which is torn off all round). On the floor of the open belly lie, between the protovertebrae, the primitive aortas, from which five pairs of vitelline arteries are given off. (From Bischoff.))

There is a great phylogenetic significance in the perfect agreement which we find between man and the anthropoid apes in these important features of embryonic circulation, and the special construction of the placenta and the umbilical cord. We must infer from it a close blood–relationship of man and the anthropomorphic apes—a common descent of them from one and the same extinct group of lower apes. Huxley's "pithecometra–principle" applies to these ontogenetic features as much as to any other morphological relations: "The differences in construction of any part of the body are less between man and the anthropoid apes than between the latter and the lower apes."

This important Huxleian law, the chief consequence of which is "the descent of man from the ape," has lately been confirmed in an interesting and unexpected way from the side of the experimental physiology of the blood. The experiments of Hans Friedenthal at Berlin have shown that human blood, mixed with the blood of lower apes, has a poisonous effect on the latter; the serum of the one destroys the blood–cells of the other. But this does not happen when human blood is mixed with that of the anthropoid ape. As we know from many other experiments that the mixture of two different kinds of blood is only possible without injury in the case of two closely related animals of the same family, we have another proof of the close blood–relationship, in the literal sense of the word, of man and the anthropoid ape.

(FIGURE 1.203. Lar or white–handed gibbon (Hylobates lar or albimanus), from the Indian mainland (From Brehm.)

FIGURE 1.204. Young orang (Satyrus orang), asleep.)

The existing anthropoid apes are only a small remnant of a large family of eastern apes (or Catarrhinae), from which man was evolved about the end of the Tertiary period. They fall into two geographical groups—the Asiatic and the African anthropoids. In each group we can distinguish two genera. The oldest of these four genera is the gibbon Hylobates, Figure 1.203); there are from eight to twelve species of it in the East Indies. I made observations of four of them during my voyage in the East Indies (1901), and had a specimen of the ash–grey gibbon (Hylobates leuciscus) living for several months in the garden of my house in Java. I have described the interesting habits of this ape (regarded by the Malays as the wild descendant of men who had lost their way) in my Malayischen Reisebriefen (chapter 11). Psychologically, he showed a good deal of resemblance to the children of my Malay hosts, with whom he played and formed a very close friendship.

(FIGURE 1.205. Wild orang (Dyssatyrus auritius). (From R. Fick and Leutemann.))

The second, larger and stronger, genus of Asiatic anthropoid ape is the orang (Satyrus); he is now found only in the islands of Borneo and Sumatra. Selenka, who has published a very thorough Study of the Development and Cranial Structure of the Anthropoid Apes (1899), distinguishes ten races of the orang, which may, however, also be regarded as "local varieties or species." They fall into two sub–genera or genera: one group, Dissatyrus (orang–bentang, Figure 1.205), is distinguished for the strength of its limbs, and the formation of very peculiar and salient cheek–pads in the elderly male; these are wanting in the other group, the ordinary orang–outang (Eusatyrus).

(FIGURE 1.206. The bald–headed chimpanzee (Anthropithecus calvus). Female. This fresh species, described by Frank Beddard in 1897 as Troglodytes calvus, differs considerably from the ordinary A. niger Figure 1.207) in the structure of the head, the colouring, and the absence of hair in parts.)

Several species have lately been distinguished in the two genera of the black African anthropoid apes (chimpanzee and gorilla). In the genus Anthropithecus (or Anthropopithecus, formerly Troglodytes), the bald–headed chimpanzee, A. calvus (Figure 1.206), and the gorilla–like A. mafuca differ very strikingly from the ordinary Anthropithecus niger (Figure 1.207), not only in the size and proportion of many parts of the body, but also in the peculiar shape of the head, especially the ears and lips, and in the hair and colour. The controversy that still continues as to whether these different forms of chimpanzee and orang are "merely local varieties" or "true species" is an idle one; as in all such disputes of classifiers there is an utter absence of clear ideas as to what a species really is.

Of the largest and most famous of all the anthropoid apes, the gorilla, Paschen has lately discovered a giant–form in the interior of the Cameroons, which seems to differ from the ordinary species (Gorilla gina Figure 1.208), not only by its unusual size and strength, but also by a special formation of the skull. This giant gorilla (Gorilla gigas, Figure 1.209) is six feet eight inches long; the span of its great arms is about nine feet; its powerful chest is twice as broad as that of a strong man.

(FIGURE 1.207. Female chimpanzee (Anthropithecus niger). (From Brehm.)

FIGURE 1.208. Female gorilla. (From Brehm.)

FIGURE 1.209. Male giant–gorilla (Gorilla gigas), from Yaunde, in the interior of the Cameroons. killed by H. Paschen, stuffed by Umlauff.)

The whole structure of this huge anthropoid ape is not merely very similar to that of man, but it is substantially the same. "The same 200 bones, arranged in the same way, form our internal skeleton; the same 300 muscles effect our movements; the same hair covers our skin; the same groups of ganglionic cells compose the ingenious mechanism of our brain; the same four–chambered heart is the central pump of our circulation." The really existing differences in the shape and size of the various parts are explained by differences in their growth, due to adaptation to different habits of life and unequal use of the various organs. This of itself proves morphologically the descent of man from the ape. We will return to the point in Chapter 2.23. But I wanted to point already to this important solution of "the question of questions," because that agreement in the formation of the embryonic membranes and in foetal circulation which I have described affords a particularly weighty proof of it. It is the more instructive as even cenogenetic structures may in certain circumstances have a high phylogenetic value. In conjunction with the other facts, it affords a striking confirmation of our biogenetic law.

CPSIA information can be obtained at www.ICGtesting.com
Printed in the USA
LVOW021648190513

334490LV00019B/787/A